SPECIAL FORCES

SURVIVAL GUIDE

A FIREFLY BOOK

Published by Firefly Books Ltd. 2014

Copyright © 2014 Marshall Editions

First printing

Publisher Cataloging-in-Publication Data (U.S.)

Stilwell, Alexander Charles, 1961–
  Special Forces survival guide : desert, arctic,
mountain, jungle, urban / Alexander Stilwell.
[192] p. : ill., col. photos. ;   cm.
Includes index.
Summary: Guide to survival skills developed for the
Special Forces, organized in to extreme environments.
ISBN-13: 978-1-77085-318-8 (pbk.)
1. Survival – Handbooks, manuals, etc.   2. Wilderness
survival – Handbooks, manuals, etc. I. Title.
613.69 dc23   GV200.5.S756   2014

**Library and Archives Canada Cataloguing in
Publication**

Stilwell, Alexander, author
       Special forces survival guide : desert, Arctic,
mountain, jungle, urban / Alexander Stilwell. Includes
index.
ISBN 978-1-77085-318-8 (pbk.)
       1. Combat survival. 2. Desert warfare. 3.
Winter warfare. 4. Mountain warfare. 5. Jungle
warfare. 6. Urban warfare. 7. Special forces (Military
science). I. Title.
U225.S75 2014   355.5'4   C2013-906538-5

Published in the United States by
Firefly Books (U.S.) Inc.
P.O. Box 1338, Ellicott Station
Buffalo, New York 14205

Published in Canada by
Firefly Books Ltd.
50 Staples Avenue, Unit 1
Richmond Hill, Ontario L4B 0A7

Printed in China

Conceived, edited, and designed by
Marshall Editions
The Old Brewery
6 Blundell Street
London N7 9BH
www.quarto.com

For Marshall Editions:
Editorial Director Sorrel Wood; **Project Editor**
Cathy Meeus; **Design and Cover Design** Paul
Turner, Stonecastle Graphics; **Illustrator** Jess Wilson;
Editorial Assistants Philippa Davis, Lucy Kingett;
Indexer Diana LeCore;
Production Manager Nikki Ingram

SPECIAL FORCES
SURVIVAL GUIDE

DESERT • ARCTIC • MOUNTAIN • JUNGLE • URBAN

Alexander Stilwell

FIREFLY BOOKS

Contents

INTRODUCTION

Special Forces are highly trained army, naval or air force personnel who take on the toughest missions. They stand apart from regular military forces because of the extent and variety of their training and their ability to master any situation and every environment.

Apart from dealing with often complex and demanding missions that require extreme skill and determination, Special Forces must also be able to survive in all types of weather. This includes polar, mountain, desert and jungle environments, with extremes of heat, cold, humidity, high winds, sandstorms, decreased oxygen, flash flooding and exposure to rapidly changing weather conditions.

In order to carry out their missions effectively — operating in the most remote areas to avoid being seen — soldiers need to endure even the most extreme conditions. For this reason, Special Forces are given very high levels of training in survival skills.

TESTED FOR SUCCESS

Special Forces are trained to learn from and respect people who are native to particular environments, whether it is the Inuit people of the Arctic, the Bedouin of the Arabian desert or the Dayaks of the Borneo jungle. Over the years, Special Forces units serving in areas such as the Middle East, Southeast Asia and the Arctic have adopted some of these native skills in order to hone their own techniques.

The methods and approaches in this book are based on tried and tested techniques used by Special Forces soldiers to survive in operational environments in different climatic areas and terrains. The priority for Special Forces is not simply basic survival but successful completion of their mission, which can mean that there are several different types of survival skills required at various stages of a single mission. For example, in the preparation and attack phase while carrying out reconnaissance or while approaching a target. In this phase, a Special Forces soldier may have limited access to supplies due to distance from resupply and will therefore need to supplement their provisions. Another typical scenario is the exfiltration or escape and evasion phase. This may be the result of Special Forces operators separating from their team after contact with the enemy or escaping from capture. Whatever the mission may be, survival and living off the land is an essential part of the Special Forces operator's range of skills.

ACTION PRIORITIES IN SURVIVAL SCENARIOS

Within the scope of this book it is not possible to cover in detail every aspect of survival in each of the environments included. However, each chapter covers key techniques to explain

▲ TRAINING TO SURVIVE

U.S. Air Force training near Kandahar, Afghanistan. All military forces receive survival training, but the survival skills of the Special Forces' elite units are second to none.

the main survival priorities — namely water, food and shelter — in each situation, organized to reflect the key survival priorities in that area.

The most urgent priorities differ between environments but are also often remarkably similar across the world's varied and challenging landscapes. The need for water, for example, is surprisingly as urgent in the polar regions as it is in the desert. However, the methods of finding it will be very different. Likewise, although the snowy environment of the Arctic suggests an obvious need for protection against the elements, the cold desert night can also cause hypothermia.

Such extreme environments demand different actions and responses to avoid the same problems. In the desert the priority for avoiding dehydration will be to minimize the effects of the sun by covering up, while in the Arctic, over-exertion in warm clothing may lead to the same end.

In all environments, in any survival situation, first aid will always be the most urgent priority. However, this book is not an exhaustive guide to first aid, so these topics are placed in the same logical order as other survival priorities.

THE IMPORTANCE OF SELF-RESCUE AND TRAINING

In all of the environments covered in this book, finding your own way to safety is often an option and occasionally a priority — such as in the mountains where it may be impossible for a rescue helicopter to land — but the methods differ. Stars always make a reliable navigational guide, but they cannot be seen as clearly from under the jungle canopy as they can in the clear skies of the desert or Arctic.

The purpose of this book is to provide a range of skills that can be adapted to different areas and circumstances, based on the hands-on experience and training of Special Forces. For Special Forces, survival field craft is an essential component of their training. In the case of U.S. Special Forces this is part of a 13-week training phase, including a five-day exercise in which the candidates test their survival skills. Once they join their units, Special Forces continue to develop their skills and experience as they are deployed in a variety of environments.

Anyone planning a trip in one of the wilderness regions of the world should get professional training in appropriate survival skills and first aid, but the tips and techniques explained in this book will give you added confidence when planning your own adventures in the exciting and challenging world of nature.

Alexander Stilwell

▲ WARMTH, LIGHT, FOOD, WATER
Special Forces are trained in a number of fire-lighting techniques. Fires provide vital warmth, light, cooking facilities and can be used to sterilize water.

◀ DIGGING FOR VICTORY
Shelter is crucial in all environments, particularly the Arctic where extreme weather conditions are a constant challenge.

How to use this book

This book is a guide to the specialist survival skills and tactics devised and deployed by the Special Forces. Using minimal tools and only basic equipment, these techniques are designed to work in the most testing environments.

Apart from dealing with often complex and demanding missions that require extreme skill and determination, Special Forces must also be able to survive in any situation. This includes regions of extreme cold, mountain, desert and jungle environments. The chapters in this book cover the specialist skills that are particular to each environment.

ORGANIZED BY CLIMATE

Each chapter includes a guide to the specific dangers unique to each environment and how to prepare in advance to avoid disaster.

A SELF-HELP MANUAL

Top priorities — shelter, water, food, making fires and getting rescued — are explained in each situation, with extra guides to first aid and even navigation methods for self-rescue scenarios.

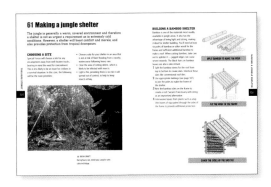

1 DESERT SURVIVAL

The desert is the setting for many of the most daring Special Forces operations in history. In conditions where you can lose up to 4 cups (1 l) of water an hour, emergency skills are not reserved for disaster situations — they are part of everyday survival in this most challenging of environments.

TAKING THE HEAT

Long-range missions in the desert demand that Special Forces learn to survive in searing heat during the day and bitter cold at night, while navigating successfully to their objectives and returning safely back to base.

Extremes of temperature are one of the chief challenges for those stranded in the desert. The heat of the desert can be punishing; during the day, the temperature may reach 122°F (50°C), but at night it can plummet to below 32°F (0°C) and the cold may be increased further by the windchill factor. To make matters worse, there is often little chance of shelter or readily accessible sources of water.

SPECIAL FORCES EXPERIENCE

In both the first and second Gulf wars, British and American Special Forces units were deployed in the desert to identify and destroy enemy missile launchers. This meant spending extended periods of time in an inhospitable environment while remaining invisible to the enemy. The ill-fated Special Air Service (SAS) unit Bravo Two Zero discovered to their cost how dangerous a mission in the desert unsupported by vehicles could be. Finding nowhere to set up a lying-up position (LUP), let alone an observation post (OP), on the hard desert floor, they were soon discovered by the enemy and some lost their lives to hypothermia while trying to escape.

◀ ON WATCH

A Special Forces operative in the desert makes use of a rocky outcrop to conceal himself.

ADAPTING TO THE CONDITIONS

Special Forces allow about two weeks to acclimatize to desert conditions, but in a survival situation gradual acclimatization isn't an option. Particular concerns include extremes of temperature, problems of moving over sandy or rocky terrain, a lack of landmarks and difficulty in finding shelter from the elements. In the desert, Special Forces stay under cover during the day and travel at night to reduce exposure to heat and minimize loss of fluids. They know that movement after dark also helps to keep the body warm in the cold of the desert night. They optimize their night vision by reducing the use of lights while on the move — a measure which is also vital for concealment on covert missions.

▲ THE LAST DROP
Strict rationing of your water supplies is essential for desert survival.

WATER IS THE KEY

The key requirement in an extreme desert environment is water. Personnel operating in the desert endeavor to carry as much water as possible or to find water sources on their route. Life expectancy in the desert is directly related to the amount of water you can drink and how efficiently you can minimize water loss from the body by reducing exposure to sun and heat.

Even with 1 gallon (4 l) of water available to drink while resting in shady conditions your life expectancy in a harsh desert environment is only three days. In order to maintain a reasonably healthy balance of water intake versus water loss in a desert with an average temperature of 95°F (35°C), you will need to drink at least 1 gallon (4 l) of water in a 24-hour period. For every hour you walk in temperatures of about 100°F (38°C), you lose about 4 cups (1 l) of water through sweating.

Food, although important in any situation, is less of a priority in the desert. In fact, because food absorbs water from the body for the process of digestion, eating may be counterproductive if water is in short supply.

SHELTERING FROM THE SUN

If you are stranded in the desert, you will not survive for very long if exposed to the heat of the sun. Likewise, it is crucial to have somewhere relatively warm to shelter when you are exposed to the opposite extreme — the severe cold of the desert night. It is extremely important to minimize the length of time you are exposed to both the heat and cold of desert conditions. One way to reduce exposure is to build a shelter, and for Special Forces this is a vital skill, not just for survival but also to provide covert bases from which to observe the enemy. Depending on the resources and equipment that you have available, different types of desert shelter can be built (see pages 28–31).

Protection from the elements can also be provided by clothing. The native people of desert regions wear long flowing robes and headdresses for a good reason. Such garments protect the skin from the direct effect of the sun's rays and shield the face and eyes from the reflected glare of the desert floor and from swirling sandstorms. Special Forces use whatever material is available, including parachutes, to wrap up warm at night.

◀ **ARAB STYLE**

This Special Forces sniper is wearing a traditional-style keffiyeh headdress to protect his face from the desert glare. Learning from indigenous people is a key Special Forces survival skill.

ON COURSE FOR SURVIVAL

In the desert, it is often sensible to be passive and wait, as even by sitting still you will be losing large amounts of water that need to be replaced.

Traveling to safety in a desert area is a formidable challenge because movement will place even more pressure on your water supplies. Navigation in the desert is also hazardous as the dangers of getting lost are potentially disastrous. This chapter emphasizes the importance of signaling for rescue as soon as possible, but also provides guidance on the best methods of finding your way in the desert, using either the sun, or — if you are traveling at night, which is preferable — the stars and the moon.

The quicker you can get to your destination, the better your chances of survival. Finding your way in the flat and often featureless desert terrain can be difficult and there is a real risk that if you get lost you may not survive. In

▲ ON COURSE FOR SURVIVAL

A U.S. unit pauses while on patrol to check their position. In the desert, you need to be certain of your route.

the desert, the ability to navigate by the stars at night is both practical — because of the large, clear skyscape and cooler temperature — and extremely useful because of the lack of landmarks. However, Special Forces also have a number of tricks up their sleeves when it comes to determining their position and direction with the aid of the sun or by following the pattern of sand dunes.

Other key survival techniques that are covered in this chapter include dealing with heat-related conditions such as heat exhaustion and heatstroke, which are a high risk in the desert.

1 Assessing your options

The desert is a punishing environment if you are unprepared for the conditions. If you find yourself in a survival situation in this climate, you need to consider how to boost your chances of survival and rescue in light of your circumstances.

YOUR PRIORITIES

Depending on your reason for being stranded in the desert, there are a number of emergency actions you should take. It is essential to take time to assess the situation, to find some shade, drink some water and make some plans. Your immediate priorities are:

FIRST AID

- Check yourself and your companions for injuries or the effects of shock and carry out first aid to the best of your abilities and knowledge.

SHELTER

- You cannot survive long in the burning heat of the sun. Try to construct an improvised shelter as soon as possible (see pages 28–31).

PLAN OF ACTION

Your main decisions when making a plan will be:

- Whether to stay where you are or move.
- What your rationing system will be for water and food.
- How to make contact with rescuers.

SALVAGE

- If your aircraft or vehicle has crashed, make sure that you salvage as much useful equipment and supplies as possible — the most important of these being any water supplies.

STAY OR GO?

When Special Forces are in need of rescue, they have to decide whether to travel to safety, to wait for rescue or to travel and try to signal to potential rescuers on route. To make this key decision, a cool-headed appraisal of the situation is necessary:

SAFETY

- Special Forces are instructed to consider the chances of being intercepted in enemy territory (less likely to be an issue for civilians in a survival situation). But regardless of this issue, the distance to safety is a vital concern.

RESOURCES

- Consider your water and food resources before deciding on action. Movement creates an extra burden on water resources and you may also need some food supplies if you attempt to travel to safety, as this will expend precious energy.

PHYSICAL CONDITION

- Do you or any of your companions have an injury or other health problem? Take into account the extreme temperatures in the desert and possible obstacles that may be too challenging for the weakest of your party.

MENTAL STATE

- Do you have the determination and mental resilience to see through what could be a long and demanding journey?

NAVIGATIONAL SKILLS AND KNOWLEDGE

- Do you have sufficient navigational equipment and skills to journey to a place where help or resources are available (see pages 42–47)? Do you have any idea of the best direction in which to travel?

EQUIPMENT

- How much equipment do you have to carry? Will you be able to transport the essential supplies you will need (see pages 40–41)?

⚠ SAFE LANDING AND TAKE OFF

A helicopter services British commandos in the desert. A safe area to land is essential for helicopter rescue — the flat desert landscape usually makes for a good landing site.

SUMMONING HELP

Special Forces usually carry a rescue beacon that sends out a signal if a team or individual goes missing. Search and Rescue (SAR) teams will then be sent out to search for them. However, the assumption on the following pages is that such electronic aids are not available.

2 Using smoke signals

If you decide to remain in the same place and wait for rescue, you can use smoke signals to indicate your position. Manufactured flares are ideal, but improvised methods can also be effective and knowing how to use them may secure your survival.

IMPROVISED SMOKE SIGNALS

During the day, you will need to create a visual sign that contrasts with your environment so that it can be seen from the air. Smoke from a fire is an ideal solution. See page 31 for advice on fire lighting.

CHOOSE YOUR MATERIALS

- The best contrast in the desert is black smoke, which can be produced by burning rubber, such as tires, or oil. Other manufactured materials may also produce dark smoke. A smokeless fire — for example, one made from dry wood — may not be noticed from the air.

EXTINGUISH IT

Beware of starting an uncontrolled fire. Be sure to fully extinguish any fire you light.

BE PREPARED

- Have everything ready to burn as soon as you spot an aircraft; they tend to appear unexpectedly and fly at high speed, so your window of opportunity may last only a few minutes. Pack dry tinder around your dark flammable materials and keep lighting equipment handy.

◀ EMERGENCY FLARE
Colored signal flares are a good option for indicating your location. To avoid attracting unwanted attention, Special Forces use them only when rescuers are close.

3 Leaving ground signals

Ground signals can be an effective alternative to smoke and flares if these are not an option in your situation. They can also provide specific information about your situation to your rescuers.

VISIBLE FROM THE AIR

For Special Forces, the advantage of ground signals is that they will not be visible to enemy ground forces. However, the need to be invisible from ground level is unlikely to be a consideration in a civilian situation. If you are in need of rescue, the most likely source of help is from the air, so use whatever dark materials you can find to create a contrast with the desert floor.

WHAT TO DO

Create a large shape such as a cross or a triangle, or construct a shape that has a recognized meaning under the international ground-to-air codes that fits your case (right).

SIGNALING AT NIGHT

In the desert, it is likely that you will be traveling at night, making you invisible to aircraft. If you are moving after dark, a way of boosting your chances of rescue is to leave ground signals at regular intervals as you move. These can be seen by rescuers during the following days and will provide a rough idea of your direction of travel and enable them to focus the area of search.

INTERNATIONAL GROUND-TO-AIR CODES

I	Serious injury
II	Need medical supplies
F	Need food and water
N	Negative
Y	Affirmative
LL	All is well
X	Unable to move
↑	Am moving this way
K	Indicate direction to proceed
⅃L	Do not understand
□	Need compass and map
△	Safe to land here
L	Need fuel and oil
⌄	Need firearms/ammunition

4 Sending mirror signals

An effective way of attracting the attention of aircraft in the desert is to send signals by reflecting sunlight. Special Forces usually carry a signaling device known as a heliograph in their emergency pack.

IMPROVISING A HELIOGRAPH

The most effective heliograph is a hand mirror, but other materials may also provide enough reflection to send a signal, including:

- The inside of a standard survival tin.
- A piece of glass.
- A strip of foil.

SAY IT IN MORSE CODE

If you know Morse code, you can also use a heliograph to send more complex messages about your condition and circumstances. The most basic Morse code signal is SOS (Save Our Souls), which indicates that you need urgent help.

MORSE CODE

Letter	Morse code	Verbal signal	Letter	Morse code	Verbal signal
A	. _	Alpha	T	_	Tango
B	_ . . .	Bravo	U	. . _	Uniform
C	_ . _ .	Charlie	V	. . . _	Victor
D	_ . .	Delta	W	. _ _	Whiskey
E	.	Echo	X	_ . . _	X-ray
F	. . _ .	Foxtrot	Y	_ . _ _	Yankee
G	_ _ .	Golf	Z	_ _ . .	Zulu
H	. . . .	Hotel			
I	. .	India	**Number**	**Morse code**	**Verbal signal**
J	. _ _ _	Juliet	1	. _ _ _ _	One
K	_ . _	Kilo	2	. . _ _ _	Two
L	. _ . .	Lima	3	. . . _ _	Three
M	_ _	Mike	4	. . . . _	Four
N	_ .	November	5	. . . . .	Five
O	_ _ _	Oscar	6	_	Six
P	. _ _ .	Papa	7	_ _ . . .	Seven
Q	_ _ . _	Quebec	8	_ _ _ . .	Eight
R	. _ .	Romeo	9	_ _ _ _ .	Nine
S	. . .	Sierra	0	_ _ _ _ _	Zero

5 Using body signals

If you see an aircraft or helicopter approaching at a low altitude, you may be able to communicate your situation and needs using a variety of body signals.

HOW TO COMMUNICATE WITH YOUR BODY

An internationally recognized range of body signals can be used to signal messages to a pilot such as where it might be safe for the aircraft to land.

NEED URGENT MEDICAL HELP

LAND HERE (INDICATE DIRECTION)

ALL OK, DO NOT WAIT

DO NOT LAND HERE

AIRCRAFT ABANDONED, PICK US UP

YES (AFFIRMATIVE)

NO (NEGATIVE)

6 Improvising desert clothing

Clothing that will protect you from the heat, cold, sun and sand of the desert can be the key to survival. In an emergency you must try to adapt whatever materials you have available to create effective desert clothing.

PROTECT AGAINST HEAT AND GLARE

Special Forces have learned that indigenous desert people such as Bedouin wear loose headdresses and clothing such as the burnoose and keffiyeh for good reason. Learning from this local wisdom, adapt your clothing for the desert with the following principles in mind:

- **Maximize insulation** — Air trapped between clothing and the body acts as insulation against heat in the daytime and also against the cold at night. Keep your clothing loose.

- **Minimize evaporation** — Despite the heat, ensure that all parts of the body are covered as you will lose more sweat by evaporation from an uncovered body than you will from sweating under clothing. This means wearing long trousers and keeping sleeves rolled down on shirts or jackets.

◀ TRADITIONAL PROTECTION

As well as shielding your face from the effects of sun and glare, a traditional keffiyeh can also keep sand and dust out of your nose and mouth.

- **Minimize exposure to glare** — Glare and heat come from below as well as above. The traditional keffiyeh headdress protects the neck and lower part of the face against glare reflected from the desert floor. If necessary, improvise with a length of spare fabric.
- **Maximize eye protection** — Shield the eyes (if sunglasses are not available) by wearing tinted goggles. Even blackening the skin below the eyes with charcoal will help to reduce the reflection of glare into the eyes.

SOAK UP SWEAT

Tie a bandanna around your neck to soak up sweat. Soak the bandanna in cool water (if sufficient water is available) to help keep your body temperature down.

LOOK AFTER YOUR FEET

It is not just the sand outside your boot that is a problem; desert sand can easily get inside your footwear and cause severe blistering and wounds while walking. Special Forces keep their feet fit for action using the following techniques in a survival situation:

1 Wrap long pieces of cloth — approximately 5 inches (13 cm) wide and 3 feet (1 m) in length — around the top of the boots to prevent sand getting inside. Do not wrap them too tightly as this will restrict circulation.

PREVENT SAND GETTING INTO BOOTS

2 Improvise a secondary layer under the boots, by tying on a piece of rubber from a tire or something similar to protect your feet from the heat of the desert floor.

◀ DESERT BOOTS

Special Forces choose boots that have sturdy soles to provide a good grip in rugged terrain and protect their feet from the heat of the desert.

7 Finding water

In the desert, finding water is a top priority and takes precedence over finding food. Ideally you need to drink about 20 cups (almost 5 l) of water per day to prevent dehydration.

MAINTAINING WATER BALANCE

In a desert you lose significant amounts of water even when resting in the shade and, in order to maintain the correct water balance in your body, you will probably need to go out in the heat of the sun during the day to find water supplies. Seeking water in such extreme heat will cause you to lose a large amount of fluid from your body through physical exertion and stress, so your survival may depend on correctly judging the risks and benefits of leaving your shelter to seek water.

PICKING LIKELY SOURCES

Special Forces try to minimize effort by focusing on the sorts of places where water is most likely to be found. The most obvious sources are rain, streams and rivers, but part of the definition of a desert is that it is waterless, so it may be necessary to find other sources as described here.

FILTER IT

Wherever you find water, you'll need to make it safe to drink by filtration and sterilization (see pages 74–75).

▲ WATERCOURSE
Water may be present in the desert, but it is often stagnant or contaminated, so must be sterilized before drinking.

PLANT SOURCES

A number of plants are excellent sources of water in the desert:

- **Cactus** — Cut off the head and either squeeze or mash the pulp to extract water. Do not eat the pulp itself.
- **Saxaul** — This is a large shrub with a spongy bark that retains water. Extract water by squeezing the bark.
- Any other desert plant that has roots near the surface may hold water. Try to extract the water by squeezing or mashing.

SAXAUL

GETTING WATER FROM DRY WATERCOURSES

If you can see traces of a dried-up watercourse, you may be able to dig down and find water still present:

1 Dig in the outside bend of a watercourse until the earth or sand becomes damp.

2 Collect any water that seeps from the surrounding earth or sand to form a pool.

MOUNTAINS AND ROCK POOLS

If you are in a rocky area, you may find water that has collected from rain in natural rock pools or in crevices and caves. Special Forces use a tube formed from a hollow grass stem to suck out the water present in a rock fissure that cannot be reached with the mouth or hand.

FOLLOW ANIMALS

Special Forces know that animals and insects can provide valuable clues to the presence of water as they are likely to be moving to or from water sources. If an animal or insect is not visible, follow footprints or droppings. Birds also often fly either to or from water. You can sometimes tell if a bird is coming from water by the pattern of its flight. If it is full of water, it may dip a bit when flying. But be aware of the dangers of sharing a water source with animals, including the following risks:

- Attack by certain animals if they are disturbed.
- Contamination of the water by their droppings. Drinking such water could lead to serious illness.

8 Making a solar still

Knowing how to construct a solar still is one of the most effective ways of producing water and can make all the difference to your survival chances if there are no obvious water sources close by.

HOW A SOLAR STILL WORKS

A solar still collects water from the earth and purifies it through a process of natural distillation. It has the advantage of involving minimum effort in daylight hours, enabling you to conserve water.

Once it has been set up, it should produce water on a regular basis — if the ground is reasonably wet, you may collect about 4 cups (1 l) of water in 24 hours — without any further physical effort, other than to move the site every two days or so.

WHAT YOU NEED

- A 6.5-foot (2 m) square piece of clear plastic sheeting.
- A watertight container.
- A 2-foot (0.5 m) length of tubing (plastic, or some other material — such as a hollow reed — if there are no manufactured materials available).

◀ STILL IN ACTION
Special Forces make a solar still in the desert. This can be an effective way of extracting moisture from the ground.

MAKE A SOLAR STILL WITH A PLASTIC SHEET

1 Find an unshaded area and dig a hole 3 feet (1 m) across and 2 feet (0.5 m) deep. At the bottom of the hole, dig a further small hole to fit the dimensions of your water container.

2 Put the container in the small hole. Place one end of your drinking tube in the container with the other end at the top of the main hole.

3 Place the plastic sheet over the hole and weigh it down with earth and stones on the edges. Allow the end of the tube to emerge under the sheeting.

4 Put a small stone in the center of the sheet, heavy enough to make it sag over the top of the container, forming a cone.

5 You can then drink the water that condenses on the undersurface of the plastic sheet and trickles down the cone into the container. Use the drinking tube or transfer it to a container to drink later.

CARRYING WATER

If you decide to travel to safety, you may not be sure of being able to find enough water on the way, and a search for water could hold you up on your journey. It is therefore important to find ways of carrying water securely.

- Screw-cap bottles are ideal containers for water but these may not always be available.
- Any form of watertight material, including animal intestine, can be used as an improvised "bladder."
- A bamboo stem can make a useful water container.

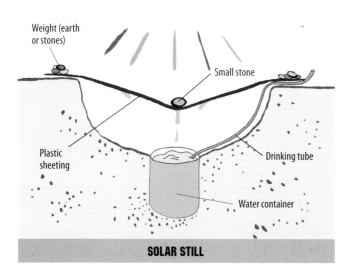

Weight (earth or stones)

Small stone

Plastic sheeting

Drinking tube

Water container

SOLAR STILL

9 Building a shelter

Finding shelter from the heat of the sun and the chill of the night is one of the key priorities for survival. In the desert, this is particularly difficult as there are few natural materials on hand, but the following techniques can help to provide protection.

SIMPLE STONE SHELTER

If you are in an area where there are plenty of loose rocks, you can use these to construct a stone shelter. In its simplest form, a roofless stone shelter serves as a windbreak and provides some shade. A more elaborate version with a roof may resemble a primitive house.

1 Build the walls from rocks in the vicinity.
2 Use grasses or branches to make a roof.

AVOID FALLING ROCKS

Do not build your shelter in an area where there are signs of rockfalls — the risk of sustaining injury is high.

NATURAL PROTECTION

As an alternative to building a shelter, you may be able to find protection in a natural rock formation such as a cave. You can add extra protection by:

- Building additional or higher walls with rocks.
- Making overhead protection from materials such as a ground cloth or parachute cloth.

TARP SHELTERS (SHOWN OPPOSITE)

If you have any form of tarpaulin, poncho liner, sacking, blankets or plastic sheets, these can be put to good use as a serviceable shelter to provide protection from the sun. You'll also need some rope or string, sticks or poles and some rocks or sandbags. Choose a method depending on the materials available.

◀ SIMPLE SHELTER
In the desert, you have to use whatever materials are available to make a shelter.

SIMPLE COVERING METHOD

1 Stretch string between two upright poles embedded in the ground.

2 Drape the fabric over the string and secure with improvised pegs.

WEIGHTED COVERING METHOD

1 Stretch string between two poles.

2 Drape the fabric over the string with one side folded under and weighted with your gear.

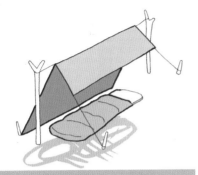

SIMPLE COVERING SHELTER

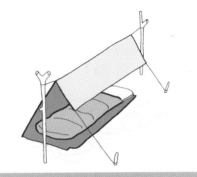

WEIGHTED COVERING SHELTER

ASYMMETRIC METHOD

1 Lash two poles together to make an A-frame.

2 Attach one end of a string or another pole to the top of the A-frame and secure the other end with an improvised peg or other means.

3 Drape the fabric over the string and weight the sides with rocks.

DOUBLE-LAYER METHOD

1 Stretch two lengths of string between two poles, one above the other.

2 Drape a piece of fabric over each string to form a double layer. Secure both layers with rocks.

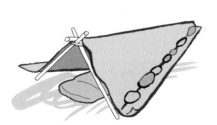

ASYMMETRIC SHELTER

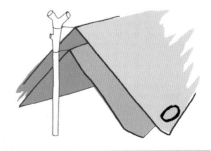

DOUBLE-LAYER SHELTER

MAKING A BELOW-GROUND SHELTER

If there are insufficient shelter-building materials readily available, you may need to dig a hideout in the sand to provide protection from the elements and boost your chances of survival.

HOW TO MAKE THE SHELTER

You will need equipment such as sandbags and ground cloths or improvised alternatives such as rocks and any available length of fabric. Building a shelter like this will significantly reduce your exposure to heat, but it will require considerable effort to build, so try to build it when conditions are cooler either early in the morning or in the evening.

1 Dig a trench about 2 feet (0.5 m) below the level of the ground with enough width and length for you to be able to lie in it with room to spare for movement and for any supplies that you want to keep with you.

2 Pile sand and rocks from the digging around the sides of the depression.

3 Place a ground cloth, or any other large piece of cloth such as a poncho liner, over the area and weigh it down around three sides with rocks or sandbags.

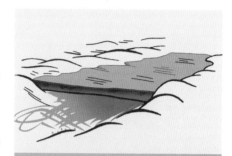

WEIGH DOWN THE GROUND CLOTH

4 Create a second layer over the first (if you have a spare ground cloth and enough materials to use as weights) to provide better insulation. Allow a gap of about 12 inches (30 cm) between the layers.

5 Weigh the second layer down around the sides as before.

CREATE A SECOND LAYER

STAYING WARM

Whenever possible, make a fire in the desert to keep warm in the often bitterly cold nights.

COLLECT FUEL

When you are moving around or traveling in the desert, it is a good idea to collect any pieces of tinder, wood, palm leaves, stems, scrub or dead vegetation that you can find and keep it for making a fire later on. Other potentially flammable materials include dry roots that you can dig out from the ground and dried camel dung.

FIRE-BUILDING TIPS

To protect your fire from the wind, build a circle of rocks or dig a small pit so that the fire is partly below the level of the ground.

ROCK PROTECTION FOR A DESERT FIRE

◀ **STONE SHELTER**
U.S. soldiers receive instruction on shelter building. This type of low-lying shelter will provide both protection from the elements and from enemy fire.

10 Finding food

Food is not the number one priority in the desert. In fact, in the desert it's important to limit your intake of food if water is very scarce. Nevertheless, if you are to maintain your energy levels over a long period, you will need to find something to eat.

FINDING FOOD IN THE DESERT

There are a number of plants in the desert that can be eaten, but it is important to be absolutely sure that you have identified the plant correctly, bearing in mind that half of all plants are either inedible or poisonous. When unsure, Special Forces perform the universal edibility test described on page 149. Another factor to bear in mind is that even if a plant is edible, it may have unpleasant side effects if eaten in large quantities. In view of this, try to vary your diet as much as possible. Some common edible desert plants are shown on these pages.

ALBA

A broom-like plant found in North Africa and the Middle East. The spring flowers are edible.

ACACIA

Found in tropical Africa and Australia and easily recognizable by its bright yellow ball-like flowers. Young leaves, flowers and pods can be cooked or eaten raw.

DESERT GOURD

Look like vines and grow in mats. Fruit and flowers can be eaten raw — seeds are also edible if roasted. Water can be extracted from the stems and shoots.

HUNTING ANIMALS

Animals are rarely seen in the desert and those that live in this environment, such as rodents or lizards, often live underground. Because of their scarcity, hunting for animals, reptiles and birds may prove to be too time-consuming and exhausting. If you are not on the move, it is best to set traps rather than attempt to hunt (see page 153).

(see page 153)

DATE PALM

This tall tree can be found from North Africa eastward to India. The fruit can be eaten.

DESERT AMARANTH

Found in many desert regions of the world. All parts of the plant are edible.

AGAVE

Native to Central America, the Caribbean and parts of Mexico. Flowers that appear on the end of long stalks can be eaten once cooked.

WILD GOURD

Mostly found in tropical and subtropical regions. Seeds can be roasted and flowers can be eaten raw. Chewing the stems and shoots usually produces some water.

11 Dealing with heat-related conditions

In the testing conditions of the desert, the most common health emergencies result from extreme heat. Knowing how to deal with these conditions can make a significant contribution to your survival chances and those of your companions.

HEAT CRAMPS

Loss of water from the body through copious sweating also leads to loss of body salts, which in turn can lead to heat cramps.

SYMPTOMS

- Sudden pain and stiffness in the arms, legs and body.

ACTION

- If possible, move the affected person to the shade or the coolest area you can find, loosen clothing and provide water to drink.

SAVED BY SALT

If you have some salt, you can make a rehydration solution to treat someone suffering from the effects of dehydration. Add ½ teaspoon (2 ml) of salt to 4 cups (1 l) of clean water. Add up to 2 tablespoons (30 ml) of sugar if available.

◀ MAINTAINING FLUID INTAKE
Special Forces know that maintaining adequate fluid intake is the key to avoiding heat-related conditions in the desert.

◀ STRETCHER HELP
Special Forces care for an injured teammate in the desert. In cases of severe heatstroke or serious injury, it may be necessary to transport a companion on a stretcher.

HEAT EXHAUSTION AND HEATSTROKE

Heat exhaustion is caused by loss of fluid through sweating, which has not been adequately replenished by drinking enough water. If unchecked, it can lead to the more serious and potentially fatal condition of heatstroke, in which the body's cooling system fails completely.

HEAT EXHAUSTION SYMPTOMS

- Increased sweating.
- Pale, clammy skin.
- Dizziness accompanied by a headache.
- Irritability and weakness.

HEATSTROKE SYMPTOMS

- Hot, dry skin.
- Nausea and vomiting.
- Loss of consciousness.

ACTION

These measures should be undertaken if you suspect either heat exhaustion or heatstroke.

1 Move the person into a cool, shaded area, preferably lying on a raised bed, platform or stretcher.

2 Loosen clothing.

3 Provide enough water to drink slowly and regularly (if the person is conscious).

4 Sprinkle the person with cool water (if there is enough available) and fan them.

5 Massage the body, arms and legs, if you suspect heatstroke.

6 Be prepared to carry the person if they lose consciousness (see how to build a stretcher, page 107).

INJURIES IN THE DESERT

Injuries of all kinds, including broken bones resulting from falls, can occur in the harsh terrain of the desert. See pages 126–127 for first aid advice.

12 Treating bites and stings

The desert can be a dangerous place not only because of its extreme environment but also because it's home to a variety of venomous insects and snakes. It's always safest to assume that any spider or snake you encounter is capable of delivering a poisonous bite.

HIGH-RISK AREAS

These creatures tend to seek protection from the extreme effects of the sun, so take care when taking refuge in caves as you may have some dangerous companions. Also be wary in any areas where there is vegetation. Use a stick to probe the area before you put your hand anywhere or lie down as you may come into contact with a venomous snake or insect.

IF YOU ARE BITTEN

If you are stung or bitten by one of these creatures, there are some techniques you can use to reduce the severity of the problem.

SCORPION STINGS

Take the following steps as soon as possible:
1 Wash the affected area with soap and water.
2 Place a cool compress or ideally ice on the area and leave it for 10 minutes. Alternatively, make a paste out of mud and ashes and apply this to the area of the sting. The following are also soothing: dandelion sap, coconut white, crushed garlic cloves, raw onion.

SPIDER BITES

It will help those that are giving you subsequent medical treatment if you can identify the species responsible for the bite.
1 Clean the area of the wound.
2 Apply a clean dressing.

CHECK YOUR BOOTS

Smaller creatures, such as scorpions, may find your clothing and equipment a good place to hide. Make sure that you turn your boots upside down and shake them before putting them on, and check your clothing and bedding as well.

SCORPION

ADDITIONAL MEASURES

In the case of a bite by a highly dangerous creature such as a black widow spider, you may need to treat the victim for shock and provide cardiopulmonary resuscitation (CPR) if you are trained to do so.
1 Lie the person down with legs raised above the level of the heart and loosen clothing.
2 Observe the person's breathing and if it stops commence CPR if you have been trained in this technique.

SCORPIONS

There are about 25 species of scorpion that are capable of delivering a poisonous sting that could endanger human life. Even the less dangerous species can inflict a painful sting.

SPIDERS

There are a number of venomous spiders that live in the desert:

- **Black widow** — The female can deliver venom 15 times as toxic as that of a rattlesnake (right). Any spider of this type should be avoided as even the less poisonous ones can cause very unpleasant side effects, including muscle pain and nausea.
- **Brown recluse spider/violin spider** — Identifiable by the violin-shaped pattern on its back, this species can deliver a potentially fatal bite.

SNAKES

Dangerous snakes that live in desert regions include the following:

- **Sand viper** — This poisonous snake is found in central Africa, the northern Sahara and the Sudan region. It has a pale body with three rows of dark spots.
- **Puff adder** — A large and extremely poisonous snake that is found in desert areas of Africa and Arabia, it is 3 to 5 feet (1–1.5 m) long. It is gray to dark brown and marked with thin yellow chevrons. Unlike most snakes, the puff adder will not always get out of your way, so be sure to avoid it.
- **Rattlesnake** — The various species of rattlesnake are found in the deserts of North America. Although they are not aggressive, they can give a painful and potentially fatal bite if they are provoked or surprised. Rattlesnakes usually have gray or light brown coloration marked with diamonds or blotches.
- **Tiger snake** — Found in desert regions of southern Australia, this snake has a potentially fatal bite. It has an olive or dark brown body with a yellowish or olive underside and is marked with cross bands.

BLACK WIDOW **BROWN RECLUSE**

SAND VIPER **PUFF ADDER** **RATTLESNAKE** **TIGER SNAKE**

13 Dealing with sandstorms

A sandstorm is a strong wind carrying a large volume of sand particles. Some sandstorms can be about 1 mile (1.5 km) high and in some cases may last for days. They can pose a serious risk to survival in the desert.

WHAT TO DO IN A SANDSTORM

If you see a sandstorm coming in time, you may be able to avoid it by getting onto higher ground such as a ridge. However, if you are caught in a sandstorm, here are some key survival dos and don'ts:

- **Do** shelter behind a large boulder or other fixed object for protection from hard flying objects in the storm.
- **Do** stay close to the ground, protecting your head with your arms or with your backpack, if there is no other protection available.
- **Do** cover your eyes, nose and mouth and as much of the rest of your body as possible.
- **Do** keep your equipment close to you — preferably attached.
- **Do** make sure that you move from side to side periodically so that you do not get covered in sand.
- **Do** remain covered until the storm has passed.
- **Don't** lie in a ditch or gully as you may be hit by a flash flood.
- **Don't** try to shelter from the sandstorm on the leeward side of a dune as this may result in you being buried in sand.

KNOW THE RISKS

Sandstorms are an ever-present danger in the desert and can be extremely dangerous. They present the following risks:

- Severely impaired visibility, making it impossible to navigate.
- Impossibility of movement because of the sand-blasting effect of the sand particles in the wind.
- Damage to the respiratory system and risk of asphyxiation from inhalation of sand particles and dust.
- Damage to the eyes from sand particles.
- Being buried in the sand.

PREPARING FOR A SANDSTORM

▲ SAND BLAST
A sandstorm can be huge and life-threatening so it is vital to know what to do to protect yourself and your team.

MAINTAINING DIRECTION

A sandstorm may well obliterate any points on the landscape that you have been using for navigation. In this respect it is similar to the effects of a snow blizzard in Arctic regions. If you have time, set out a marker such as a stick in the ground that you can use to help re-establish your direction.

QUICKSAND DANGER

Be aware that after a sandstorm there may be an increase in the size and number of areas of quicksand — tread carefully.

SANDSTORM EFFECTS

The khamsin wind in the Sahara is a notable cause of sandstorms in that area, where the sandstorm is known as the *simoom* or "poison wind." One of the added dangers of the *simoom* is the raised temperature, which can reach 130°F (55°C). This can cause heatstroke and contribute to dehydration.

SANDSTORM IN AFRICA

14 Travel light

When traveling in the desert, you need to aim to travel light. Any extra weight that you have to carry places further demands on your physical resources and will result in a greater need for water that may be in short supply.

YOUR BAGGAGE ALLOWANCE

In the desert you should aim to carry no more than 35 pounds (16 kg). Your survival may depend on allocating a significant part of your weight allowance to carrying water. A gallon of water weighs 10 pounds (4.5 kg) so that would leave about 25 pounds (11.5 kg) for other items. Depending on what resources you have available, your equipment may include:

- Shelter-making materials, such as a tarpaulin or a parachute canopy.
- A compass and map.
- Knife.
- Salt.
- First aid kit.
- Torch.
- Stick or pole, which can be used as a walking aid and as a support for a shelter.
- Minimal food rations.

HOW TO CARRY YOUR GEAR

When traveling, you will need to consider how you will carry your equipment. If you do not have any store-bought packs, you may have to improvise. Special Forces use options such as those described opposite to carry gear in an emergency.

IMPROVISING A BACKPACK

You can make an improvised backpack using a spare piece of fabric gathered at the top to form a bag. Attach this to a frame constructed from any pieces of wood or bamboo that may be available. You will need some form of binding material to hold it all together and a length of rope or material to form the straps. There are two basic frame design options:

SQUARE FRAME

1 Place two sticks or poles of the intended height of your frame parallel to each other.
2 Lash several horizontal slats securely between the two sticks to create a frame.
3 Attach lengths of rope between the top and bottom of the frame at each side to form the carrying straps.

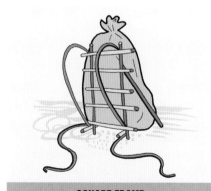

SQUARE FRAME

WISHBONE FRAME

1 Lash two sticks or poles together at the top to create a wishbone shape.

2 Lash several horizontal slats between the arms of the "wishbone" to create the frame.

3 Attach lengths of rope between the top of the frame and to the bottom at each side to form the carrying straps.

WISHBONE FRAME

⚠ **MOVING LIGHTLY**

For speed and flexibility, U.S. Special Forces on a desert operation are carrying minimal equipment.

BLANKET BANDOLIER

A bandolier for carrying small supply items can be created quite simply from a blanket or a ground cloth and some string or rope.

1 Wrap your equipment in the blanket or ground cloth.

2 Tie the ends.

3 Put it over one shoulder and around the waist and secure it with string or rope.

WRAP YOUR EQUIPMENT

CARRY YOUR BANDOLIER

15 Choosing a route

If rescue is not possible or likely, and you know where to aim for, finding your own way to safety may be your best option. If you decide to travel in daylight, knowing how to ascertain your position and the direction in which to travel according to the sun will be critical.

NAVIGATING BY SHADOWS

Shadows move in the opposite direction to the sun. In the northern hemisphere they move from west to east and at noon they indicate north. In the southern hemisphere, they move from east to west and at noon indicate south. You can use this knowledge to judge the position of north, south, east and west. Here are two methods used by Special Forces. You'll need two sticks and, for the intersecting arcs method, a piece of string.

JOINING SHADOW MARKS

1 Take a stick about 3 feet (1 m) in length. Push the stick upright into level ground when the sun is shining. Mark the end of the shadow cast by the stick with a rock or other marker.

2 Allow 15 to 20 minutes for the shadow to move and mark the second position of the shadow.

3 Then draw a line on the ground through the two shadow points. This line indicates the direction from east to west.

4 If you now place your left foot close to the first point and your right foot close to the second point, in the northern hemisphere you will be facing north, and in the southern hemisphere, south.

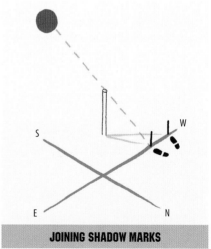

JOINING SHADOW MARKS

▲ SHADOW SIGN
A stick creates a shadow on the sand that can be used for finding direction.

INTERSECTING ARCS

1 Follow Step 1 of the previous method. Once you have indicated the first shadow mark, tie a piece of string joining the central stick and marker stick. Using the joined sticks, make an arc through the first shadow mark and all the way round it.

2 Wait for the shadow to reduce in length as the sun rises. Once the sun has reached its highest point, the shadow will lengthen again, at a different point to where it previously fell.

3 Mark the point where the new shadow touches the arc that you have drawn. The two points you have marked will be your east/west line.

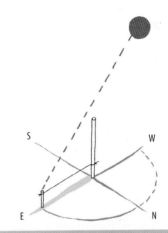

INTERSECTING ARCS

FOLLOWING NATURAL FEATURES

Special Forces benefit from specialized understanding of the physical features that may help or hinder a journey across desert terrain. Here are two rules they follow:

OBSERVING THE TERRAIN

KEEP TO RIDGES
By following a ridge you will get a better view of the lay of the land and find it easier to navigate. Obviously, the ridge you are on may not necessarily be heading in the direction that you want to go so you may need to make some adjustments and compromises — for example, leaving a ridge and picking up another one later on.

FOLLOW WATERCOURSES
Natural watercourses can be useful routes for travel. However, be aware that these are prone to flash floods — being caught by one of these can be fatal.

16 Navigation at night

The clear skies of the desert allow for relatively simple navigation by the stars. The night skies in the northern and southern hemispheres require different knowledge, but having discovered north or south, you can then plan your route to safety.

NORTHERN HEMISPHERE STAR NAVIGATION

The central point in the northern hemisphere is the North Star (also known as Polaris or the Pole Star). As the North Star itself can be comparatively dim, it helps to check its position against the nearby constellations: the Big Dipper (also called the Plough), the Little Dipper and Cassiopeia.

FINDING THE NORTH STAR

1 The Big Dipper is made up of seven stars. Follow a line upward through the two stars at the end of the "dipper" to reach the North Star, which is at the end of the handle of the Little Dipper constellation.

2 Cassiopeia is formed by five stars roughly in the shape of a "W," one side of which is comparatively flat. By bisecting the angle formed by the two stars of the flat side of the "W" you will produce a line that points directly at the North Star.

ONCE YOU HAVE FOUND THE NORTH STAR

The North Star is a useful navigational tool because it is always located almost directly above the North Pole and is therefore a reliable guide. It is positioned at the same vertical angle above the horizon as the northern line of latitude where you are standing.

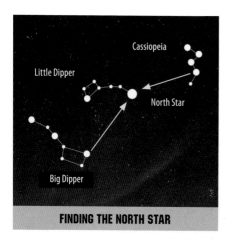

Cassiopeia

Little Dipper

North Star

Big Dipper

FINDING THE NORTH STAR

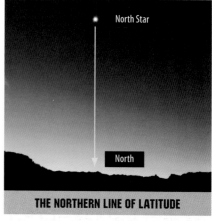

North Star

North

THE NORTHERN LINE OF LATITUDE

MAKING A COMPASS

If you do not have a compass, it is possible to make one from a thin length of metal such as a needle.

MAKE A MAGNET

1 To magnetize the metal, stroke the end of it with a magnet or on silk.

2 As you are unlikely to have a magnet or silk in a survival scenario, you can produce a similar effect by stroking the metal across a piece of leather or even through your hair.

FROM MAGNET TO COMPASS

1 Once the metal has been magnetized, suspend it from a piece of string or either an animal or human hair. Suspending the piece of metal inside a transparent bottle will make the "compass" less likely to be affected by air movement.

2 The piece of metal will align itself in a north/ south direction.

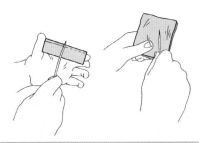

MAGNETIZING A METAL NEEDLE

MAGNETIZED METAL HUNG INSIDE A BOTTLE

◀ **NIGHT TRAVEL**
Special Forces often prefer to travel at night in the desert.

SOUTHERN HEMISPHERE STAR NAVIGATION

The southern hemisphere does not have a single conveniently located star over the South Pole but the constellation known as the Southern Cross provides a useful guide for navigation. It consists of four stars in a roughly cross-like arrangement.

HOW TO LOCATE DUE SOUTH

1 Pass a line through the longer part of the cross, through some faint stars in a constellation beneath and into an area sometimes known as the Coal Sack where there are no stars.

2 Take an angle bisecting the two stars located to the left of the Southern Cross (known as the Pointer Stars) and it should intersect with the line you have drawn from the Southern Cross to indicate a point that indicates south.

3 Identify a landmark beneath the southern point you have worked out using this method and use that as a guide.

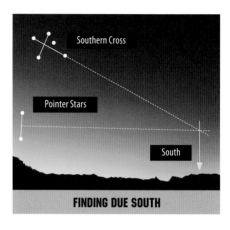

FINDING DUE SOUTH

▼ TRAVELING AFTER DARK
U.S. troops return to their shelter as night falls. Once the sun has set, the moon can be a vital aid to direction-finding.

A Special Forces soldier
scanning the ground for
enemy movement.

NAVIGATING BY THE MOON

The moon orbits the Earth and, as it does
so, reflects the sun from different angles. If
the moon rises before the sun has set, the
illuminated side of the moon is its western
side. If the moon rises after midnight, the
illuminated side is its eastern side.

QUARTER-MOON NAVIGATION

You can also use quarter or sickle moons
to give you a rough indication of compass
directions.

1 Draw a line through the horns (top
and bottom points) of the quarter or
sickle moon.

2 Note the point where the line touches the
horizon. This is roughly south, if you are in
the northern hemisphere, or north, if you are
in the southern hemisphere.

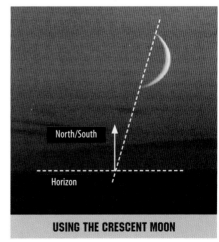

North/South

Horizon

USING THE CRESCENT MOON

2 EXTREME COLD

Survival in regions of extreme cold presents unique and serious challenges. In the Arctic and other polar regions, where temperatures often drop to −40°F (−40°C), the pressing need is for warmth and shelter. In these conditions, Special Forces techniques can make all the difference to your chances of survival.

LIFE IN THE FREEZER

Freezing temperatures – often exaggerated by the effects of severe windchill – snow- and ice-covered terrain and the risk of snowstorms and blizzards are the key features of this harsh environment that can present huge risks for survival.

EXTREME COLD

Extreme cold regions of the world are those in which the temperature rarely, if ever, rises above freezing. The main regions that fall into this category are the northern and southern polar regions (Arctic and Antarctica) and also the mainly treeless environment, also known as the tundra, that includes much of the area north of the Arctic Circle. The areas above the tree line on mountains, especially in winter, are also extremely cold.

The Arctic region consists of a frozen ocean, while Antarctica comprises a landmass that is between 10,000 and 13,000 feet (3,000–4,000 m) above sea level. It is almost entirely covered by ice, which can be up to 10,000 feet (3,000 m) thick. In both regions the climate is extreme, particularly in winter. Summers are warmer, but even then the warmth of the sun is reduced by the reflective effect of ice and snow. In the ice cap areas, the temperature never rises above freezing, 32°F (0°C), and in the tundra it never exceeds 50°F (10°C).

In midwinter there may be total and constant darkness, while midsummer brings 24 hours of daylight. Winter in the Arctic starts in late August and continues until the end of December, whereas in Antarctica it lasts from March to September. Temperatures are lowest during winter, but another important factor so far as survival is concerned is the windchill factor. A 20-mile-per-hour (32 km/h) wind can reduce a −4°F (−20°C) temperature to −30°F (−34°C).

▲ **SNOW WALKING**
A serviceman uses snowshoes on an exercise. With the right equipment and knowledge you can survive extreme cold.

SPECIAL PREPARATIONS

To acquire the skills they need to survive in these conditions, Special Forces train in countries on the edge of the Arctic Circle, such as Norway. They are aware that in whatever extreme environments they may find themselves, the inhabitants of these areas not only survive but often thrive. Special Forces know that part of the challenge of survival is adaptation and learning from the shelter-building and hunting techniques used by the indigenous peoples.

This preparation equips personnel for deployment in operations as varied as dealing with rescue missions in civil emergencies in the Antarctic region to reconnaissance and direct military action in the harsh winter conditions of countries such as Afghanistan.

Using the knowledge gained from this experience, and taking advantage of the greater individual choice allowed in Special Forces that is not permitted among ordinary service personnel, Special Forces operatives will often take items they have selected themselves — for example, specialist clothing layers and footwear, made from high-specification materials — over regular military issue equipment.

SURVIVAL ESSENTIALS

In extreme cold, effective shelters are vital to maintain a minimum of warmth and to provide protection against windchill, blizzards and so on. You are likely to confront the difficult decision of whether to continue your journey to safety, or take shelter and wait for an improvement in the severe conditions. Sufficiently warm clothing can make the difference between life and death, but

51

knowing how to make best use of the clothing materials at hand can also be critical.

Adequate food intake is also essential for maintaining body heat and resilience to the effects of cold. Food resources are minimal in both polar regions, but if you have no supplies, you can hunt seals and birds, and catch fish. In the tundra, there is plenty of vegetation — mostly consisting of moss, lichen, herbs and small shrubs. There are also animals, including foxes, wolves, hares and caribou (or reindeer),

and birds such as the ptarmigan and snow goose. Fish are often plentiful in the streams and rivers of these regions and provide a ready source of food.

Water is always vital — perhaps surprisingly in regions of extreme cold. But even moderate activity in warm clothing can cause copious sweating and there is therefore an ever-present risk of dehydration.

MENTAL RESILIENCE

Morale and attitude are also key factors in determining the ability to survive. Cold can have a seriously depleting effect on your energy levels and can reduce your will to take the necessary action to overcome the challenges of this environment. It requires extra mental effort to take the essential steps

for survival, such as building shelters and making fires, in conditions of extreme cold. Survival in extreme cold depends ultimately on your courage, determination and knowledge of how to cope.

BE PREPARED, AVOID PANIC

This chapter covers some of the essential skills used by Special Forces in extreme cold environments, including careful choice of clothing, finding food and water, building shelters, navigation and traveling to safety and other ways of seeking rescue. It also describes how to make best use of the equipment you may have with you, along with local resources you may find, to help you to survive. The focus is on survival measures for Arctic regions but it is applicable to any area where extreme winter weather may strike and create snow-bound and icy conditions.

The guidance and skills in this chapter will help you to cope with a variety of challenges. The greatest enemies of survival are panic, confusion and lack of action. Once you have a plan and use the skills shown here, you have a much better chance of surviving. Special Forces withstand extreme environments by working with the elements and not fighting them.

◀ TIMBER SHELTER
Shelters are essential for survival in extreme cold environments. Intensive training helps Special Forces not only to survive but also to operate efficiently.

▼ SNOW HOME
An igloo-style shelter provides excellent protection from the elements, but may be time-consuming to construct.

▲ WELCOME SIGHT
A helicopter may be your best chance of rescue in the polar regions.

17 Time to shelter

Building even a rudimentary shelter as protection against icy winds could be a lifesaving action, and is your priority if you find yourself stranded in an extreme cold environment. Do this before considering other actions such as seeking rescue.

SECONDS COUNT

Special Forces caught unexpectedly in harsh winter conditions will prioritize shelter building. If they are moving from one location to another and a blizzard strikes, they will make a rapid decision about whether it is quicker to return to a previously built shelter or to build a new one on the spot. In this situation, consider the following questions:

- How long will it take you to build a shelter, taking into account where you are and what tools and other resources you have? If it is going to take you longer to build a shelter with the resources you have available than it will take to return to your previous base, it is likely to be better to return.

IN BLIZZARD CONDITIONS

Navigation becomes almost impossible in blizzard conditions when you are likely to become disorientated. In general the best advice is to seek shelter until the blizzard has subsided. However, depending on the severity of the blizzard, the navigation equipment you have available, and how urgent it is for you to continue, you may decide to keep going, although this is a risky course of action.

- Is it a whiteout? If so, there is a real danger that you will not be able to navigate to your previous shelter and that you will get lost. Even if you manage to navigate to the right area, your shelter may have been covered in snow and you will not be able to find it.
- Is there potential for building a shelter in your current location? If not, it may be better to keep moving.

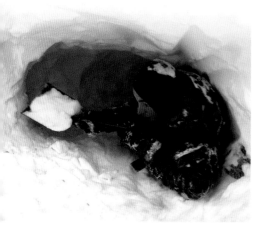

◀ DIGGING IN

A U.S. Marine digs a snow cave. In blizzard conditions a temporary shelter such as this can save your life.

18 Making a shelter from snow

When conditions are extreme and your energy levels may be low, the shelters described on these pages can be relatively quick to construct and provide adequate temporary shelter in an emergency.

SNOW BLOCK SHELTER

In an emergency, you can quickly build a windbreak from blocks of snow, which will provide some protection from the windchill. If you have a ground cloth or other large piece of material, you can use this to add a roof to this basic shelter, which will provide greater protection from the elements.

1 Secure one end of a ground cloth or other large piece of material between two layers of snow blocks.

2 Anchor the other end with a single layer of snow blocks.

HOW TO CUT BLOCKS

To cut snow blocks you need an edged tool such as a large knife, a shovel or an ice ax. The blocks should be about 18 x 20 inches (45 x 50 cm) and 4 to 8 inches (10–20 cm) thick. You may need to test the snow to ensure the blocks will be firm enough for use. You cannot make blocks from powdery or loose snow.

3 Place something dry, such as your backpack, on the floor of the shelter to sit on.

4 Remain in the shelter until the blizzard has passed and it is safe to emerge.

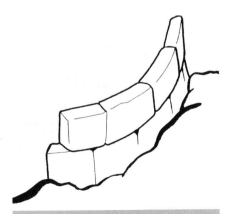

BUILD A WINDBREAK

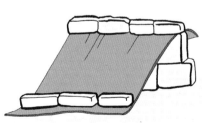

ADD A ROOF

MAKING A SNOW TRENCH

This type of simple shelter will provide refuge from the worst of the weather for a short period of time. It requires a tool such as a saw, shovel or machete for cutting snow.

1 Mark out the area of the trench based on the amount of space you need to lie down comfortably, plus some room for equipment and with allowance for movement.

2 Dig the trench to a depth of about 2 feet (60 cm). Remove the snow in blocks equivalent in width to your trench. The blocks should be about 8 inches (20 cm) thick.

3 Cut a ledge into the sides of the trench, about 6 inches (15 cm) in width and height.

4 Place two blocks opposite one another on the ledges either side of your trench. Lean the two blocks together so that they meet in the center above the trench creating a roof.

5 Place a block to close the windward end of the trench.

6 Use a removable block or sheeting to form a door at the other end of the trench.

▼ **PROTECTIVE TRENCH**

In this hollow dug from the snow, this U.S. Marine is protected from windchill as well as enemy fire.

MAKING A SNOW HIVE

This shelter is similar in shape to an igloo, but is simpler to construct. You will need sheeting and foliage (or similar materials). You should be able to build this shelter with your hands (wearing gloves), but a shovel will be helpful.

1 Place the foliage on the sheeting and create a ball large enough to create a space in which you can lie down, sit up, and store your equipment.

2 Pack snow around the ball to a thickness of about 12 inches (30 cm), leaving a gap for an entrance.

3 Make sure that the opening is large enough to provide sufficient air for ventilation, and that the hive is not liable to collapse.

4 Once you are sure that the snow is firm, gradually remove the contents of the foliage ball through the entrance, leaving an empty space inside.

PACK AROUND SNOW

CHECK OPENING

REMOVE CONTENTS

SHELTER FAST

Special Forces know that weather conditions in cold regions can change very quickly. Knowing how to construct a hasty shelter can be vital. A simple trench with a tarpaulin roof is the quickest shelter to build and may save you in an emergency.

19 Making a shelter in wooded areas

In a wooded area, the trees themselves will provide some added protection. Your choice of shelter will also be influenced by how much energy and time you have, and the tools available.

FALLEN-TREE SHELTER

If you are in a forest and come across a fallen tree, you may find that this provides an ideal basis for a shelter that is relatively quick to construct.

1 Cut away excess branches "inside" the shelter to increase the space available.

2 Dig out any snow underneath the natural tent shape of the tree.

3 Use any branches you have cut to line the shelter for added insulation.

▲ SURPRISINGLY SNUG
Conifers provide excellent natural protection that can be enhanced by digging a pit around the base of the trunk.

▶ FIR-TREE DEN
A rudimentary A-frame supports the conifer branches that make up this simple but effective shelter.

TREE-PIT SHELTER

If there are no fallen trees, you can take advantage of the natural protection provided by the branches of an upright tree. This is an ideal shelter to build in a forest and it is also relatively economical in terms of both time and energy expenditure.

1 Find a tree with a natural hollow area around the base and dig the snow out of this area to create a comfortable area for yourself and your equipment.

2 Use the snow you have dug out to create a windbreak around your shelter.

3 Any branches above you will provide a natural roof. Add to this protection by cutting branches from the other side of the tree and placing them on top of the branches over your shelter. You can also use branches to insulate the sides and floor of the shelter.

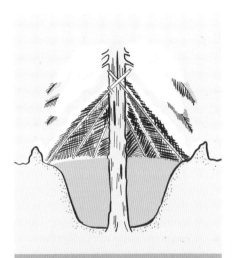

TREE-PIT SHELTER IN CROSS-SECTION

20 Getting rescued

In conditions of extreme cold, getting rescued usually provides your best chance of survival. Once you have built a shelter, you'll need to give urgent consideration to how you can attract attention from potential rescuers.

SIGNAL YOUR POSITION

One way to signal your position is to dig trenches in the snow to create a shadow that will be visible from the air.

1 Dig a trench about 3 feet (1 m) wide in a north–south direction.

2 As you dig, pile up the snow along the edges of the trench to increase the shadow effect. Aim to make the trench about 30 feet (9 m) in length.

3 If the snow is not too deep, you may be able to reveal the dark earth beneath, which will provide ideal contrast. Otherwise, line the trench with any vegetation you can find nearby.

SAFE LANDING

If you want a helicopter or aircraft to land in the area to effect a rescue, you need to be sure that the area is safe for landing.

- Check how waterlogged the ground is, whether there are any areas of thin ice or any major obstructions that would make landing too dangerous.
- Mark a safe landing zone for a helicopter with a large "H," which you can create by digging trenches (as described at left).
- Mark a safe landing area for an aircraft with a large triangular shape. In the snow, you can dig three trenches to make a triangle.

MAKE A WIND SOCK

It will be helpful for the pilot of any rescue helicopter to know the strength and direction of the wind at the landing site. You can help by hanging spare clothing on a stick to make a wind sock.

◀ HERE I AM
Dark clothing makes this survivor clearly visible against the snow-white background.

MAXIMIZING THE CHANCES OF A SUCCESSFUL RESCUE

Remember when preparing for rescue that the safety of the rescue team is paramount. The pilot will not risk their rescue aircraft and crew if they consider the situation to be too dangerous. The recognized order of priority is the aircraft, the crew and winchman and then the survivor.

To maximize the chances of a safe and successful rescue by helicopter, follow these guidelines:

1 If you are advising the rescue team to land their aircraft, make sure that you have established a secure landing area.

2 Check that the ground is firm and not waterlogged.

3 Check that there are no hidden obstacles, holes, boulders or tree stumps, or any loose branches that could interfere with the landing or prevent an aircraft from taking off.

4 Make sure that the likely approach of the helicopter is not obstructed by any hidden hazards such as power lines.

5 Check that there are no severe side winds and, if possible, set up a wind sock or a fire with smoke so that the pilot can assess the wind direction and strength.

NOTE: To convey information on the landing site to the pilot of the rescuing aircraft, see the information on signaling with your body on page 21.

▶ **RESCUE FROM THE COLD**
Special Forces have extensive support for any mission into polar regions to ensure any injured parties can be rescued quickly.

21 Layer clothes to survive

In extreme cold weather conditions, having the right clothing is critical – it can mean the difference between life and death. Special Forces will generally follow what is now well known as the "layering principle."

WHAT IS LAYERING?

Even in extreme cold environments, the human body can generate considerable heat and moisture when moving around. The layering principle aims to reduce the moisture that is created by sweating, which can contribute to exposure and chilling. These are the essential four layers:

1 Wicking layer An undershirt and leggings made from a technical fabric that keeps the body warm without absorbing moisture.

2 Intermediate warm layer A long-sleeved garment made from wool (ideally merino wool), which also has good wicking properties and is breathable and comfortable to wear.

3 Thermal layer A jacket such as a fleece or smock, with either a full or half zipper, or a thermal jacket. This may be paired with thermal pants.

4 Weatherproof layer An outer shell consisting of a waterproof and windproof jacket and pants made from a waterproof and breathable material such as Gore-Tex.

HOW TO IMPROVISE LAYERS

If you do not have designated clothing for extreme cold conditions, you will need to improvise with whatever materials you have available:

- **Salvaged equipment and materials** A parachute provides an excellent source of material for clothing. Stuffing from vehicle seats may also be useful for added insulation between clothing layers.
- **Animal skins** If you have success in hunting animals such as caribou or seals, strip the skin and clean it thoroughly. Dry it before wearing it fur-side in to maximize insulation.
- **Plants** If available, use dried plant fibers (leaves or grasses) as added insulation.

▲ PROTECTIVE LAYERS
The layering system enables you to adapt the amount of clothing you wear to your level of activity.

THE WATCHWORD IS "COLDER"

The acronym "colder" can be used as a way of remembering the principles that will help you to maintain your body temperature in conditions of extreme cold.

C — CLEAN CLOTHING

By keeping your clothing clean you maintain its efficiency and insulating properties. Cleanliness also helps to ward off infections.

O — OVERHEATING

This should be avoided in order to prevent excess sweating. If you sweat and your clothes become wet, you can quickly get chilled. To regulate your body temperature, loosen your clothes at the neck and wrists and either remove or adjust your headgear.

L — LOOSE LAYERS

The layering principle allows warm air to be trapped between the different layers of clothes and loose clothing allows your body circulation to work efficiently. Layering also allows you to remove excess clothing as necessary to adjust your body temperature according to the conditions and your levels of activity.

D — DRY

It is important to keep clothing as dry as possible. This means both avoiding the soaking effect of sweating and protecting your clothing from rain and snow.

If clothing does become wet, dry it as best you can by placing it near a fire. Another option is to take advantage of freezing temperatures by leaving the wet clothing outside to freeze. You can then beat the clothing so that the ice falls away, leaving the clothing dry.

E — EXAMINE

Regularly check your clothing for dirt, damp or damage.

R — REPAIR

Repair any tears or holes in clothing as soon as possible before they get worse. Special Forces often carry a small pack of needles and thread with them.

LAYERED FOR ACTION

22 Protect your extremities

Your head, hands and feet can have an enormous influence on both your real and perceived sense of comfort. There is nothing like dry, warm feet for making you feel more positive, as well as being essential in preventing frostbite and other disabling conditions.

KEEP YOUR HEAD WARM

About 47 percent of heat loss is through the head. Headwear is therefore vitally important in extreme cold, for instance a wool-lined cap with ear muffs. Special Forces may also be issued with a woolen head-cover. According to your level of exertion, you may need to remove headwear to cool the body. Loosening clothing at the neck can have the same effect.

PROTECT YOUR HANDS

Hand protection is vital in cold weather and layering can help to maintain warmth. Ideally, the inner layer should be a pair of thin woolen gloves and the outer layer a pair of waterproof mittens. Improvise these layers with whatever materials or spare clothing is available.

▶ COVER UP

A British Royal Marine wears a hat to maintain body heat and to protect his ears. A "face-over" provides protection from frostbite, and sunglasses shield the eyes from both direct and indirect glare from the sun.

HOW TO WATERPROOF YOUR FOOTWEAR

If you are wearing improvised footwear or if your boots are not waterproof, tie on an extra waterproof layer.

1 Cut a triangle of waterproof material (for example from a ground cloth). Use more than one layer if you have enough material.

2 Pull the ends up over your foot.

3 Tie the ends around your ankle or make holes in the points of the triangle and lace the material around your ankle.

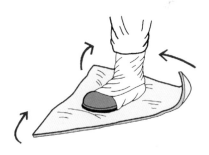

WRAP YOUR FEET

DRY YOUR FEET

Dry feet can massively boost your sense of well-being and prevent a range of foot conditions that could reduce your mobility. Here are some tips from the Special Forces:

- Dry wet socks by a fire if possible.
- Wrap damp socks around the waist to use body heat to speed drying.
- Dry wet boots by warming — but not overheating — a stone by the fire and placing it inside the boot. This will help to dry the boot from the inside while you dry them from the outside by placing them near, but not too close, to the fire.

LOOK AFTER YOUR FEET

Special Forces are normally issued with wet-weather boots that have extra layers of insulation to protect against the cold and are typically designed for temperatures down to −4°F (−20°C). The boots typically include a waterproof inner "bootie" made from technical fabric. Special Forces try to follow the principles listed below for footwear in extreme cold. If possible, improvise with available materials to create similar foot protection.

- Wear thin socks with high wicking properties next to the skin and thicker wool and technical fabric socks on top.
- Try to avoid constricting blood flow by wearing overtight boots or too many socks.
- Aim to keep the boot uppers dry and prevent water getting in over the top of the boots. For this reason, gaiters or mukluks — warm, felt-lined, flat-soled boots — are often worn by Special Forces.

23 Dealing with the cold

The most obvious health issues in extreme cold environments are hypothermia, frostbite and injuries that may result from these conditions. Your survival will depend on prevention by keeping as warm as possible and on taking prompt action if you notice the first signs of trouble.

HYPOTHERMIA

Dangerous loss of body heat that can be fatal, hypothermia is one of the principal risks of an extreme cold environment. This results from long exposure to low temperatures or from an incident such as falling through ice into cold water. Always be on the lookout for the signs of hypothermia in yourself and your companions and be ready to take action.

SYMPTOMS

Always assume hypothermia is likely if someone falls into cold water and act accordingly. In other cases, suspect hypothermia if you notice any of the following in yourself or a companion:

- Protracted and intense shivering that becomes uncontrollable.
- Difficulty speaking.
- Slow and irrational thoughts.

◀ STRETCHER TO SAFETY
U.S. Marines carry an "injured" teammate on a stretcher as part of an exercise. Injuries in extreme cold environments require fast coordinated action to get the victim to shelter and safety.

◀ VITAL TRAINING

U.S. Marines wade through ice and water to reach a waiting helicopter for evacuation. Extreme training of this kind may save lives in a real survival scenario.

ACTION

- Warm the body as soon as possible, but gradually, in a warm sleeping bag or other covering. Another person should lie alongside to transmit their body heat to the affected person.
- Put on dry clothes if possible.
- Give hot, sweet drinks if available and if the person is fully conscious.

FROSTBITE

This is damage to body tissues from exposure to extreme cold. A mild case of frostbite may only affect the skin. However, protracted exposure to the cold can lead to deeper frostbite, which can eventually lead to the loss of extremities such as the toes. The same advice as for hypothermia applies to the prevention and treatment of this condition. Do not wait for symptoms to occur before taking action.

SYMPTOMS

- Loss of feeling in hands, feet, toes or parts of the face.
- Pale skin in the affected area.

ACTION

- Keep all extremities and exposed areas as warm and dry as possible.
- Keep socks and boots clean and dry.
- Wear adequate head and face covering.
- Maintain circulation by moving toes, fingers and facial muscles.

ACT FAST

Special Forces are selected partly for their quick reactions, and are trained to move fast when required. If you fall into icy water, it is important to act quickly to prevent hypothermia by immediately taking the measures described on this page.

24 Starting a fire

A fire provides a huge boost to your survival chances in cold conditions. It will keep you warm and dry, and enable you to cook and boil water. It also provides a vital lift to morale.

FIRE-MAKING STAGES

You'll need different materials for each stage of fire building, from tinder to make the first flames to fuel for feeding and maintaining a fire large enough to provide warmth and heat for cooking. Collect flammable materials as you encounter them when moving through cold and barren terrain. For further advice on different types of fire, see pages 70–71.

TINDER

Start a fire with light and flammable dry material such as birch bark, dry grass, pine needles, bird down or wood shavings.

KINDLING

Once you have the first flames, feed your fire with twigs and small sticks of wood from which you have cut away any damp parts.

FUEL

When flames are established, maintain your fire with larger pieces of dry wood. You can also use peat and coal if you can find it.

MAKING FLAMES

Special Forces normally have windproof matches in their survival kit, but there is always a risk that these may run out. As a backup, the kit usually includes a fire-lighting device, consisting of a piece of rough steel and a small saw. Pulling the saw across the steel creates sparks which can be used to start a fire. There are other ways of creating flames, including the use of the hand drill method (right).

STARTING A FIRE USING BIRCH BARK

MAKING FLAMES THROUGH FRICTION

This method works on the principle that friction eventually produces heat and then a flame. A hardwood base is essential and this method can take a long time to produce enough heat to create a flame.

1 Find a hardwood base and cut a v-shaped notch. Create a drill stick from a round stick of medium-hard wood (for example, conifer) about ¼ inch (6 mm) in diameter.

2 Make a hole of the same diameter as the stick in the base at the point of the notch.

3 Roll the stick fast between your hands, moving your hands gradually downward, so that the stick stays in the hole you have made and the friction increases.

4 Eventually, the end of the stick should start to glow red. Place some dry tinder nearby so that you can blow gently on the red-hot end of the drill to create your first flames.

5 As the fire takes hold, gradually add kindling and then larger pieces of wood to build the flame and the fire.

USING FRICTION TO START A FIRE

◀ PREPARING FOR FIRE LIGHTING

A soldier sharpens a drill stick to be used for fire making on a survival course at a specialist training center.

25 Types of fire

To get optimum benefit from your fire, you need to build the right type of fire for the amount of fuel available, the rate that it is likely to burn, the length of time you will stay in that location and whether you need it more for warmth, cooking or signaling.

WHEN FUEL IS SHORT

Build a star-shaped fire if you do not have much wood. This type of fire can be used for keeping you warm and for cooking or boiling water.

1 Arrange the fuel wood in a star pattern radiating out from the center. Place kindling at the center and start the fire.

2 As the fire burns, push logs in toward the center to keep the fire going. Alternatively, if you want to conserve wood while you are away from the area, you can remove the logs from the center and rekindle the fire when you return.

▶ **LIGHT THE TINDER**
A Special Forces operator encourages flames by blowing on a newly ignited fire.

FIRE FOR SIGNALING

Build a tepee (or pyramid) fire if you have plenty of fuel available and you need heat rapidly. It is also ideal for signaling. If you choose to have this kind of fire outside in an extreme cold environment, create a windbreak so that the fire is not overwhelmed by the weather.

1 Place tinder in the middle of the fire.

2 Push a stick into the ground so that it is leaning over the tinder.

3 Lean small pieces of kindling against the slanted stick, leaving an opening in the direction of the wind to allow air into the fire.

4 Protect the flame from the wind while you are lighting it by shielding it with your back.

5 Once the fire has taken hold, gradually add longer bits of kindling and wood.

FIRE FOR COOKING

This type of fire, sometimes called a Yukon stove, is useful if you are going to be in one place for a relatively long period. It is protected against bad weather and is good for cooking. The other advantage is that you can construct this stove inside a shelter for added warmth. If building this kind of fire inside a shelter, it is essential to allow air to flow through your shelter to remove the dangerous carbon monoxide that the fire creates.

1 Dig a pit 6 inches (15 cm) deep and place your tinder, kindling and fuel wood in it.

2 Construct a shell around the fire from mud and stones, leaving a hole in the center as a chimney, and a gap on one side to allow air in to feed the fire.

3 Balance a cooking pot or a roasting spit over the stone surround or place wrapped food in the entrance at the base of the oven.

PYRAMID FIRE

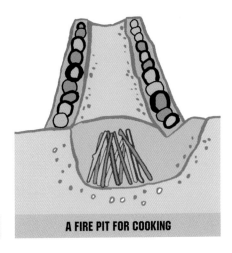

A FIRE PIT FOR COOKING

26 Obtaining water for drinking

In an extreme cold-weather environment, finding water is not usually a big problem. The challenge is more likely to be converting water from ice, snow or swamps into water that is safe to drink.

SOURCES OF WATER

In polar and tundra regions you can usually obtain water from the following sources:

- Blue (fresh) glacier water.
- Melted ice or snow.
- Brown surface water (in tundra areas).

All the water you collect should be treated with water sterilization tablets or boiled. Brown surface water should also be filtered (see pages 74–75).

HOW TO MELT ICE OR SNOW FOR DRINKING

Choose from the following methods, depending on your circumstances:

BODY HEAT

Melt the ice or snow in a waterproof bag under your clothing (but not next to your skin). This is a useful technique if you are on the move, but may take time and may also reduce body heat, so avoid it if you are at risk of hypothermia.

▶ FILLING UP
U.S. airmen on a survival exercise fill plastic water bags with snow. Their own body heat will melt the snow but they will need to purify it with tablets or boil it before drinking.

BAG METHOD

Pack the ice or snow inside a bag that will allow water to drip into another container and hang it near a fire.

MELTING SNOW INSIDE A BAG

FIRE AND STONE

Place a lump of ice or snow on a sloping stone set over a fire. Position a heatproof container at the end underneath the snow to catch the water running off.

Alternatively you could use the heat of the sun (see right).

NOT THIRSTY?

Even if you do not feel thirsty, you still need to drink at least 8 cups (2 l) of water each day to avoid dehydration. Make sure you drink regularly. A headache or drowsiness could be a sign of dehydration.

MELTING SNOW WITH SUNLIGHT

One way to melt snow for drinking water is by using the heat of the sun. You will need a container, some plastic sheeting and some way of supporting the sheeting at an angle, such as some sticks or piles of stones. If you do not have a plastic sheet, you can use a large flat piece of stone, angling it toward your container.

1 Fix the plastic sheeting at an angle toward the sun, and place a container at the lower end to catch the water.

2 When the plastic sheeting has been warmed by the sun, sprinkle snow lightly on the sheeting and wait for it to melt and run into the container. Continue the process until you have enough water.

MELTING SNOW WITH SUNLIGHT

27 Making water safe to drink

Although your main concern may be to avoid dehydration, use of contaminated water leading to an infection that causes diarrhea and vomiting could put your survival at risk. Having collected water, be sure to treat it to remove germs.

REMOVING THE RISK

It is vital to treat any water that you drink, especially from areas where the water is still, including melted snow and ice. Water from a fast-flowing stream running over rocks may be clean enough to drink without treatment, but it is best to be safe; if you drink unclean water to satisfy your thirst in the short term, you may suffer a debilitating sickness in the long term.

1 Remove soil and visible contaminants by filtering water from ponds, rivers or swamps (see opposite).

2 Kill invisible germs by using water sterilization tablets or potassium permanganate crystals (if you have these).

3 If you have no purifying additives, boil the water, preferably in a covered container, for at least 10 minutes before allowing it to cool and drinking it.

⚠ **BOILING HOT**
Boiling is a simple way of eliminating harmful germs from water. Use it to make a comforting hot drink to warm you up and boost your morale.

AVOIDING DEHYDRATION

The body can easily become dehydrated in cold conditions, partly because you are less likely to feel thirsty than in warm conditions and therefore drink less, and partly because layers of warm clothing can cause heavy sweating, which increases the risk of dehydration. Another possible problem is cold diuresis, which causes to you to lose more water through urination than normal.

Ideally, you should aim to drink at least 8 cups (2 l) of water per day, and more if you are doing strenuous physical activity.

MAKING A WATER FILTER

An improvised water filter removes particles of soil and other visible contaminants from water. To make one you will need a porous cloth bag such as a sock, or even panty hose, as well as a container.

1 Place layers of stones, followed by sand, then stones again in layers for about two thirds of the bag. Other materials you can use for filtering include moss, fresh grass and charcoal.

2 Place a container beneath the filter and then pour your water into the top and catch the filtered water at the bottom.

3 Once filtered, sterilize the water as described opposite.

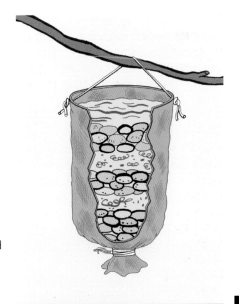

LAYERS OF STONES AND SAND

◀ **CLEAN AND FLOWING**
Water that flows over clean rocks in a mountain stream is more likely to be clean enough to drink than slow moving water farther downstream. However, it is always safest to treat the water, by boiling or by the use of water sterilization tablets.

28 Fishing under ice

If you find yourself in a survival situation in an area of frozen lakes or rivers, polar regions or frozen sea, the possibility of fishing under the ice offers a vital lifeline to an often-plentiful food resource.

CHOOSE YOUR TECHNIQUE

There are two essential ice-fishing techniques — a hook and line dropped through a hole in the ice and the use of a net underneath the ice. Choose between them according to the materials at hand and the time available. Always make sure that the ice will bear your weight before attempting to fish.

HOOK AND LINE ICE FISHING

This is the most straightforward fishing method and it means that you can simply sit and wait for a fish to bite. You will need a hook (a bent pin can serve this purpose) and a line (improvise with thread or fine string).

1 Choose a suitable area of ice and cut a hole about 12 inches (30 cm) wide.

2 Attach a weight to your line, along with the hook. Bait it and lower it until you can feel the weight hitting the bottom.

3 Once you feel the weight hitting the bottom, pull it up about 12 inches (30 cm) and jerk it up and down a bit to attract the fish.

4 Secure the line to a stick placed across the hole. This gives you the option of adding another line later.

5 Attach some form of device that will signal any movement on the line.

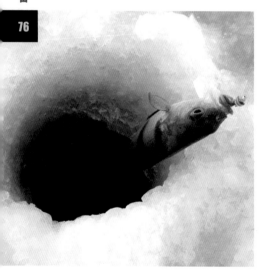

▲ FISHING IN THE FREEZER

Fish are one of the most abundant food sources in Arctic regions and can be caught with a bait and hook through an ice hole, with preparation and patience.

CUTTING THROUGH THE ICE

If the ice is thick enough to bear your weight safely, it will be difficult to penetrate, so you will need a sturdy tool to be able to cut through the ice — use a heavy knife or an ax.

NET FISHING UNDER THE ICE SHEET

Net fishing can be an efficient way of catching large quantities of fish, but it is also a relatively complicated procedure that requires time and some skill to set up.

1 Find a pole and attach a sheet of netting to it at one corner of the top of the net. Attach a line to the other top corner of the net. (See page 79 for instructions on how to make a net.)

2 Weight the bottom of the net so that it will hang open under the ice.

3 Cut two or even three holes in the ice, spaced at about 1 foot (30 cm) less than the length of the pole.

4 With the free end of the net secured to sticks or a heavy object, pass the pole, with the net attached to it, through the first hole.

5 If you have cut three holes, use the second hole to guide the pole and net toward the third hole.

6 When the end of the pole reaches the final hole, pull it out through the hole, leaving the net behind, and secure the net, so that it hangs vertically under the water.

REMOVE THE POLE FROM THE WATER

29 Knots for fishing

Knowing how to tie the correct knot can make the difference between crucial nourishment and starvation in a region where fish are the only food resource. Familiarity with tying specialist fishing knots can help you to improvise when you have few materials.

ATTACHING A FISHING HOOK

Use this knot, known as a turle knot, to attach a hook with an eye to a fishing line.

1 Thread the line through the eye of the hook.
2 Make an overhand knot (see page 183).
3 Form a bight (loop) and pass it through the overhand knot to form a slipknot.
4 Pull the knot tightly around the shank of the hook.

MAKING A LONGER FISHING LINE

If you need to join two lengths of fishing line to make a line that is long enough for your needs, use this knot.

1 Overlap the two lines from opposite directions.
2 Make five turns of one line around the other.
3 Bring the end of the line you have used for wrapping back between the two lines.
4 Repeat the wrapping process with the other line.
5 Pull the lines slowly in opposite directions.

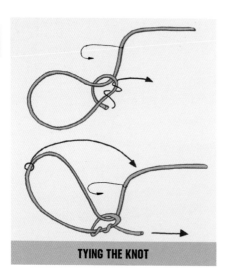

TYING THE KNOT

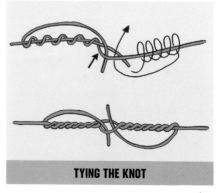

TYING THE KNOT

THE FINISHED KNOT

THE FINISHED KNOT

MAKING A FISHING NET

If you do not have a net for fishing but you do have some long pieces of cord, you can make your own net. Special Forces may have access to lines from a parachute.

1 Tie a line horizontally between two trees or posts.

2 Make a loop in one end of a second line. This line should be six times the intended depth of your net in length.

3 Wrap the loop around the horizontal line and bring both ends through the center of the loop three times.

4 Attach further lines to the horizontal line spaced about 1 inch (2.5 cm) apart.

5 To create the netting, tie adjacent lines together with an overhand knot (see page 183) until you reach the end. Make further rows until you have created a net of sufficient depth for your needs.

6 Once the main part of the net is finished, you can thread a casing rope along the bottom to give the net more strength and to add weight to the lower edge.

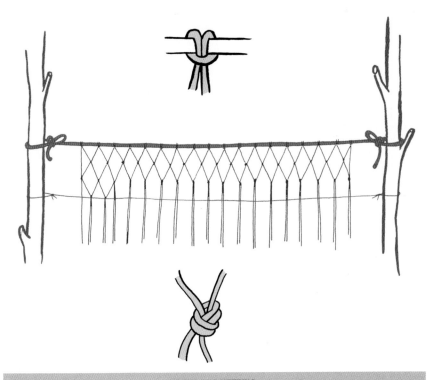

TYING THE NETTING

30 Eating your catch

It's an achievement to catch a fish, but knowing how to prepare your fish for eating – with or without fire – once you have landed your catch is also a vital survival skill.

CLEANING A FISH

Whether or not you intend to cook your fish, you will need to clean and gut it before eating.

1 Place a sharp knife under the jaw and cut down to the tail.

2 Remove the fish innards.

3 Wash out the body cavity with clean water.

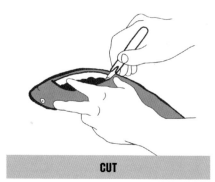

CUT

REMOVE GUTS

CLEAN

PRESERVING YOUR CATCH

An advantage of an extremely cold environment is that you have a ready-made freezer for storing food when you have more than you need for future use. To freeze your catch, gut the fish and cut it into strips. Place it to freeze outside your shelter, out of reach of potential scavengers.

HOW TO PREPARE FROZEN FISH

If cooking over a fire is not an option and you need an alternative way of preparing and eating fish, you can try a method used by Arctic dwellers. It is a quick and easy method of getting the nutrition you need. Raw fish is also said to produce more internal body heat during the digestive process than cooked fish does.

1 Clean the fish and remove all the bones.

2 Cut the fish into slices and lay them on an area of clean ice.

3 Wait for the fish to freeze, then eat.

◀ **FROZEN FISH**
A large catch of fish from a successful fishing session. One of the advantages of freezing conditions is that it is possible to preserve excess food for longer than in warmer climates.

COOKING FISH ON A FIRE

To cook a fish, you'll need to use the fire-building skills explained on pages 68–71. When you have caught a fish, it should be cooked (the exception being the frozen method described above).

■ Cut off the head, unless you are planning to place the whole fish on a spit.

■ To cook the fish, spear it with a stick supported or held over the fire, turning it so that it doesn't burn, or simply wrap it in leaves and place it directly on the embers.

COOKING FISH ON A SPIT

31 Hunting and catching animals

There are a number of animals in extreme cold regions that can be either hunted or trapped for food. Hunting or trapping animals such as reindeer or rabbits requires similar skills to those used in any other environment (see also pages 152–153).

HOW TO CATCH A SEAL

Seals can be caught when they come to the surface to breathe through a hole in the ice.

1 Wait by the hole for a seal to come up for air and then club it with a rock or any available heavy implement.

2 Quickly drag the seal out onto the ice. Larger seals should be caught and clubbed when they are on the ice as they are likely to be too heavy to haul out onto the ice from the water.

3 Strip the skin and then remove the innards.

4 Cook and eat immediately or freeze the meat to cook and eat later.

TRAPPING ARCTIC BIRDS

A few species of birds remain in the Arctic regions during the winter months, including ptarmigans, snow chickens/partridges, grouse, jays and owls. The rock ptarmigan and willow ptarmigan can be approached relatively easily and can be caught using a snare tied to the end of a long stick. You can also collect and eat birds' eggs wherever you can find them.

BEWARE OF BEARS

In the Arctic, you will need to take care that another predator such as a polar bear is not attracted by your kill. Be alert and prepare to abandon the animal you have caught rather than confront a dangerous rival.

▷ WHITE GAME
A ptarmigan in its white winter plumage. This common Arctic game bird can easily be snared with a lasso snare if you approach it stealthily.

SNARING ANIMALS IN THE SNOW

Hunting can be a laborious and potentially dangerous way of finding food and depends largely on having the right weapons. Snaring animals for food, by contrast, is comparatively straightforward, requires less energy, is less dangerous and, once the snare is set, all you have to do is wait.

SETTING A RABBIT SNARE

To set a snare for a small animal such as a rabbit, you need to try to identify their likely runs so that you can place the snare in the best place. Try to minimize as far as possible any trace of your presence. In extreme cold environments, your footprints will be visible but if there is a snowfall, your tracks and any other disturbance of the ground and snow will be covered up. To set a drag snare:

1 Confirm the presence of an animal run by signs such as disturbed ground or animal droppings.

2 Push a stake into the ground on the run.

3 Tie a noose to the stake so that the animal's head will pass freely through it. Set the bottom edge of the noose about 4 inches (10 cm) from the ground and make the diameter of the noose roughly equivalent to that of two clenched fists.

MAKING A SNARE

If you don't have a store-bought snare, you can construct one from a 30-inch (80 cm) length of wire. Make a small loop in one end and then pass the other end through the loop to create the noose. Attach this to a length of nylon or twine.

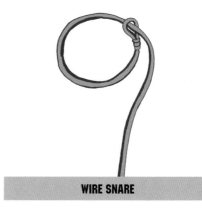

WIRE SNARE

▷ **ARCTIC HARE**
Hares and other small animals can be found in Arctic regions and are best caught by using carefully laid snares.

32 Plants you can eat

Near the poles there are few plants, but you can find plenty where the snow cover is not permanent, providing valuable nutrients, including vitamin C. Knowing which ones can be safely eaten could dramatically boost your chances of survival.

KNOW WHAT TO EAT

The edible plants described on these pages are all found in cold environments and could aid your survival. If you are uncertain whether it is safe to eat a plant you have found, use the edibility test (page 149).

GRAINS AND SEEDS

In subarctic climates you may find edible grains and seeds. Try them raw, but if they are too bitter, dry roast them in a pan over a fire.

CLOUDBERRY

This looks like a bramble and has palm-shaped leaves and a white flower. Eat the berries raw.

ARCTIC WILLOW

This is a shrub with rounded leaves. You can eat the young shoots and leaves, as well as the peeled roots.

BILBERRY

Found in the northern tundra, this is a low-growing shrub with small green leaves. Eat the berries raw. They are a good source of vitamin C.

ICELAND MOSS

This is a bushy lichen which can either be brown or gray-white. It grows in the northern polar region, providing food for both animals, such as caribou and moose, and humans. Soak it for several hours then boil thoroughly. Eat it as a vegetable or add it in dried form to soups or cereals.

BEARBERRY

This evergreen shrub is found in the far north of Europe, Asia and North America. It generally occurs as ground cover with bronze foliage in winter and white, pink or pink-tipped flowers in spring. The whole plant can be eaten if cooked thoroughly. The berries are also edible.

REINDEER MOSS

This green lichen is found widely in the northern polar region. Prepare it as per Iceland moss.

VITAMIN RICH

Bilberries and bearberries are excellent sources of vitamin C, a nutrient essential for good health. Take advantage of these precious plant foods whenever you find them.

BEARBERRIES AND BILBERRIES

33 Moving on snow

You may decide that your survival chances will be improved by walking to a place where rescue is more likely. Moving on snow for long distances demands specialized walking techniques, and you will need to take precautions against snow blindness.

WALKING TECHNIQUE

In regions of extreme cold, it is important to take into account the likely weather conditions before attempting to travel and also to assess the terrain. If you are walking on snow, you may need to make improvised snowshoes (see page 88).

- Take a sturdy travel pole with you (improvise by cutting a tree branch if necessary). This can be used as an aid for walking, a shelter support or a cooking stick.
- Follow a regular walking rhythm so that you do not become exhausted and dehydrated.
- Stop at least every hour and give yourself a 10-minute break.
- Give yourself plenty of time to build a shelter before it gets dark.

◀ DECKED OUT
This soldier has modern professional equipment for walking on snow, including snowshoes and ski poles. In a survival situation, you may need to make your own snowshoes and improvise walking poles.

◄ **IN THE SHADE**
These U.S. Marines on exercise in snow-covered terrain all wear sunglasses to prevent snow blindness.

PREVENTING SNOW BLINDNESS

Snow and ice reflect the ultraviolet rays of the sun into the face and this can cause disabling eye symptoms. These effects usually subside if protective measures are taken.

SYMPTOMS

- Severe eye pain.
- Red, watery eyes.
- Itchy eyes that may feel as if they have dust in them.
- Headache.

PREVENTION AND TREATMENT

- Wear sunglasses or goggles. If you do not have any, try improvising (described right).
- Blacken the area of skin under the eyes with some charcoal if you do not have goggles or there is no possibility of improvising. This will reduce the ultraviolet rays that reflect from the skin into the eye area.
- Bandage the eyes until symptoms subside if symptoms are severe.

IMPROVISED EYE PROTECTION

If you have no sunglasses or goggles, you can screen your eyes with a piece of transparent, colored plastic, if available, or make improvised goggles as described here.

1 Cut a slit or slits in a piece of material — for example cardboard, thin wood or tree bark.
2 Attach a length of string at each side.
3 Place the strip over your eyes and secure it at the back of the head.

IMPROVISED GOGGLES

34 Making snowshoes

Deep snow can be a serious problem when walking. You will need snowshoes to prevent your feet from sinking deep into the snow if you are to travel any distance or if you need to move around to hunt and set traps.

WHAT YOU NEED

To make improvised snowshoes you will need to find some flexible sticks to create the outer frame. This may be a green branch from a tree. You will also need some more rigid sticks for the inner supports. The method you choose depends on the materials available.

HOOP METHOD

For each shoe, you'll need two long, flexible sticks and several shorter ones, some rope, string or strips of fabric and some pieces of card or thin wood.

1 Bend a stick to form a large hoop and tie the two ends together.

2 Cut more flexible sticks and tie them across the hoop that you have created in the areas where they will provide the most support for your boot.

3 Tie pieces of card or fabric in a crisscross style (like a tennis racket) down and across the hoop to help distribute weight.

4 Use the rope or other binding material to tie the snowshoe to your boot.

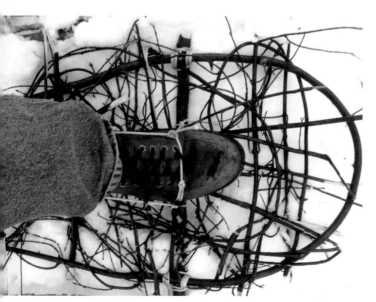

◀ IMPROVISED SHOES
This improvised snowshoe is made using the hoop method, in which a pliable branch is bent round and tied at both ends, in a tennis-racket shape.

CROSS-STICKS METHOD

For each shoe, you'll need five straight, flexible sticks and several shorter, rigid ones and some rope, string or strips of fabric.

1 Cut five straight sticks to a length of about 3 feet (1 m) and tie all the ends together to make a point.

2 Bind the sticks together at the other end, leaving a gap of about 1 inch (2.5 cm) between them.

3 Attach two or three strong sticks across the other sticks in the central area where your boot will be placed.

4 Tie string to the pointed front and pull it back, creating a prow effect, before tying the other end of string to the boot support nearest the front.

5 Place your boot on the structure and lash it securely to the snowshoe.

TIE ENDS TOGETHER IN A POINT

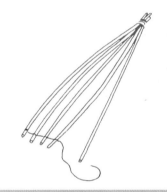

BIND THE OTHER ENDS TOGETHER

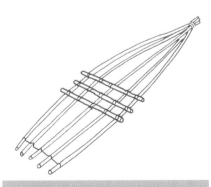

ATTACH STICKS ACROSS CENTER

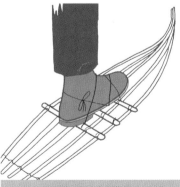

LASH YOUR BOOT IN PLACE

35 Moving on ice

There are some advantages to moving on ice as it can provide a stable walking platform in contrast to thick snow or boggy ground. However, you can never trust ice and you must always be extremely wary of it giving way, placing you in extreme danger.

WALKING ON A FROZEN RIVER

Falling through ice is potentially fatal, first because you may be trapped under the ice, and second because it is difficult to pull yourself out again. If you choose to walk on a frozen river, follow these precautions:

- Do not attempt to walk on a frozen river or lake in spring or summer. Ice is likely to be much thinner in summer months than in the winter, especially at low altitudes.

- Always stay close to the bank so that you can move quickly to firm ground if the ice starts to crack.

- Stay roped together, if you are moving in a team, and make sure that each member carries a stick, which can be used to test the ice.

▲ ROPED TOGETHER
A Special Forces team crosses a glacier. They are roped together in case anyone should fall into a crevasse. The same considerations apply to walking on a frozen river.

STICK IT OUT

When walking on ice, hold a long stick horizontally, so that if you suddenly fall through the ice, the stick will bridge the hole and enable you to pull yourself out.

A LONG STICK CAN SAVE YOU

▲ SECONDS COUNT
A U.S. Navy SEAL gets firsthand experience falling through ice during a survival exercise. This type of specialist training can make the difference between life and death in a real emergency.

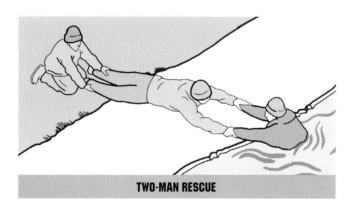

TWO-MAN RESCUE

HOW TO HELP

If you see someone fall through the ice, do not attempt to walk on to the ice to help them as you may fall in as well or make the surface even less able to support the victim. Try to reach the person with a stick or tree branch. Throw a rope, if available, to the person. If you have time, tie a loop in the rope so that they can put it around their shoulders. Either pull directly or use an anchor to secure your end of the rope. If you are with a companion, you can try a two-man rescue. One person holds the other's feet, as he moves on his stomach toward the victim and grabs his arms.

WHAT TO DO IF YOU FALL THROUGH ICE

If you are unlucky and fall through the ice, the priority is to pull yourself out as fast as possible. You can lose your life within minutes of falling into freezing water. Try to stay calm and use the following method to get yourself out of danger:

1 If you are underwater, try to maintain visual contact with the hole. If the ice has snow over it, the hole will appear darker than the surroundings. If the ice has no snow over it, the hole will appear lighter.

2 Once your head is above water, get rid of anything that might be holding you back, such as a backpack.

3 Keep your breathing as calm and as deep as possible, countering the reflex to hyperventilate caused by the cold.

4 Find the strongest and thickest part of the ice to pull yourself out on.

5 Use either the stick you've been carrying, which, with luck, has become lodged across the top of the hole you have fallen through, or stab a sharp tool such as a knife (ideally

one in each hand) into the ice outside the hole and use it to pull yourself out.

6 Get as much of your body out as possible, then kick your feet as if you were swimming, keeping your body as horizontal as possible.

7 Once you are out, keep yourself flat on the ice and head to the bank, preferably by the way you came.

8 Remove your wet clothes as soon as possible and build a fire to warm yourself (see also Hypothermia, page 66). If you are in a team, your companions can help build the fire and lend you some dry clothes. It may take a few hours for your temperature to return to normal, so you may need to make a temporary shelter to rest in while you recover and your clothes and boots are drying.

SELF-RESCUE — USE KNIVES TO GET OUT

◀ **BETTER OUT THAN IN**
A Canadian soldier digs into the ice to drag himself out of a frozen lake during winter warfare training.

36 Finding your way in polar regions

If you decide to walk to safety, you'll need to be able to travel toward your destination. Although clear skies may make navigation by the stars (see pages 44–47) easier in polar regions, the landscape is often featureless, making it difficult to find landmarks to guide you.

CLUES TO DIRECTION

It is a bonus to have a compass in a survival situation, but if you don't have a store-bought compass, you may need to improvise (see page 45) or use the sun as a guide (see pages 42–43). Natural features can also provide valuable clues to direction and your location. You can use these alongside other knowledge you may have of the area you are in to plan the best route toward safety.

CLOUDS

Clouds will appear black when over open water, forests or ground that is free from snow and ice. They appear white when over ice or snow, and are mottled when over pack ice or drifting snow.

RIVERS

Rivers flow toward the sea. This can provide a valuable insight into your position. Follow rivers downstream to travel in the direction of the coast.

BIRDS

Sea birds tend to fly out to sea in the morning and back toward the land at night. Learn from the activity of birds to orient yourself toward the sea if there are no rivers to provide a route.

MOVING LANDSCAPE

- Icebergs can be unreliable as a guide to position and direction; they often move over time.
- The clear polar air can make the distance to the location you are aiming for seem shorter than it really is, which may lead you to underestimate the time needed to reach it.

ESTIMATE YOUR POSITION

In order to estimate your current position and direction of travel as you make your journey to safety, try to sketch out a map of each leg of your journey, including any notable features, and also keep a log. If you are lucky enough to have a map or a compass, you will be able to judge your position fairly accurately, but even a rough map and improvised methods of direction finding will improve your chances of finding your way out of danger.

3 MOUNTAIN SURVIVAL

Survival at high altitude in terrain that is by definition rugged, precipitous and often treacherous can be a massive challenge. Special Forces receive in-depth training for these conditions to enable them to complete operations in mountainous regions throughout the world.

SURVIVING MOUNTAIN MISSIONS

Craggy peaks, perilous cliff faces and slopes of loose scree are all part of the challenge of a mountain environment. In areas when rescue may be out of the question, Special Forces need to use skill and ingenuity to make their way to safety.

Mountain belts may be hundreds of miles wide and thousands of miles long, with individual peaks often connected by sharp ridges and separated by deep valleys. Major mountain systems of the world include the South American Andes, the North American Cordillera — which stretches from Alaska to Mexico — the Alps in Europe and the Himalayas in Asia.

CHANGEABLE CONDITIONS
The shape of mountains and the high altitude has a radical effect on mountain climate, and this can have a significant impact on your survival strategies. Temperatures fall by about 1°F (0.5°C) for every 320 feet (100 m) gained in height. Mountains also force winds to rise and cool. There is often heavy rain or snow on the windward side of a mountain, whereas the leeward slopes tend to be drier.

One of the chief problems for anyone stranded in a mountainous region is that the weather can be very difficult to predict and is likely to change quickly from calm conditions to stormy and from warmth to extreme cold

within a relatively short period of time. There can also be radical differences in weather between relatively close locations.

Almost all mountainous regions have a similar type of vegetation above the tree line. In these areas, sparse plant life such as grasses and shrubs predominate. Such vegetation and the presence of certain animal species offers the possibility of food for those in a survival situation in this environment.

◀ ALTITUDE APTITUDE
Mountain sickness is best avoided by a slow ascent.

PHYSICAL CHALLENGES

Apart from the weather, one of the main physical challenges to be faced at high altitudes is a reduction in oxygen levels. This can have a significantly detrimental effect on both mental alertness and physical capability.

Dehydration is also an issue in mountains because a substantial amount of body moisture can be lost through rapid breathing (which may be caused by the low levels of oxygen) and by exertion, from strenuous activities such as traveling up a steep mountainside. Another contributory factor is that mountain air is drier than air at lower altitudes.

However, the main physical challenge is perhaps that presented by the terrain itself, including the dangers posed by hidden gullies and crevasses that may lead to falls and consequent serious injuries. There is also the possibility of avalanches of either stones or snow, slipping on wet rocks and the cold-related risks of frostbite and hypothermia.

SPECIAL TRAINING AND EQUIPMENT

Special Forces are highly trained in mountain warfare. Mountains provide unique opportunities for covert activities and observation and can also provide a launchpad for effective missions. However, they are also some of the most demanding environments in which to operate and survive.

Knowing how to travel safely is a key survival skill in this type of mountainous environment, where negotiating natural obstacles can present huge risks to those with

▶ **REMOTE SURVIVAL**
U.S. Marines use survival
skills in the mountains of
Afghanistan searching for
caves and hiding spots used
by the Taliban.

little experience of the conditions. Special
Forces receive extensive training in mountain
operations and in any team there is usually an
operator with advanced mountaineering skills.
Mountain training includes instruction for cliff
climbing in coastal regions, which enables
Special Forces to carry out reconnaissance and
assault missions. This high level of preparation
has enabled Special Forces to perform
successfully in a range of environments, such
as the mountainous regions of Afghanistan.

When embarking on a mountain mission,
Special Forces pay particular attention to
the clothing and equipment that they carry
with them. Clothing principles are likely to
be very similar to those discussed in chapter
two, Extreme Cold. These principles include
the layering system and the importance of

regulating clothing according to the ambient
temperature, weather conditions and level
of activity. Mountains can demand very high
levels of physical exertion so the need to
regulate body temperature to reduce water
loss through excessive sweating should always
be a high priority.

Special Forces also understand
the importance of wearing specialist
mountaineering footwear, usually of synthetic
materials. Depending on the specific mission,
Special Forces may also carry a variety of
specialized climbing gear such as crampons,
helmets, ropes and harnesses. They will always
aim to use the best equipment available, and as
the most advanced climbing gear is developed
for recreational and competitive use, they
will often use this high-level mountaineering

equipment rather than standard military issue apparatus for such missions.

LEARNING FROM EXPERIENCE

Other major issues facing Special Forces deployed in mountains range from building mountain shelters in steep and snow-bound areas, to anticipating and avoiding avalanches and moving safely in challenging mountainous terrain. Knowing how to improvise expert climbing techniques, such as belaying and rappeling, when specialist mountaineering gear is not available, can often be key to survival. Special Forces are also trained in the essential skill of arresting a fall.

THE ESSENTIALS OF LIFE

While keeping warm, building a shelter and finding your way to safety may be immediate priorities for those stranded in a mountain

region, finding water and food are nevertheless essential to survival for any length of time. The opportunities and techniques for securing these vital prerequisites for life are much the same as those covered in chapter two, Extreme Cold. Another key aspect of securing the best chance of overcoming mountain conditions is knowing what to do if something goes wrong — for example, if someone is injured. Knowing what to do in cases of mountain (altitude) sickness and basic first aid skills for fall-related injuries, using improvised supplies if necessary, are part of the Special Forces armory and are crucial in emergency situations.

▲ ESSENTIAL SUPPLIES
Climbing gear is specially chosen by individual Special Forces operatives.

▲ ROPE WORKOUT
Marines are trained in rappeling during Mountain Warfare Training.

37 Getting rescued

Rescue from a mountain is always fraught with difficulty.
The altitude and terrain may mean that it is too dangerous
for a helicopter to land. Your best option is to try to get to a
place where safe landing is a possibility.

RESCUE OPTIONS

On a mountain, self-rescue is the ideal course
of action, and you should descend to a lower
level if possible. However, this may not be an
option — for example if you or a member of
your group is injured. In this case, you need
to do whatever you can to make the job of
rescue teams easier. This may mean making a
stretcher so that you can transport an injured
companion to a safe landing site.

IMPROVING YOUR CHANCES

To improve their chances of being rescued,
Special Forces are aware of the methods
used by Search and Rescue (SAR) teams. In
a mountainous area, helicopter search and
rescue teams are most likely to use the contour
pattern, which involves flying in ascending
concentric circles around the mountain.

Special Forces may be equipped with search
beacons, but if the beacon isn't working or
they do not have a beacon, they make their
position clear to the rescuers by one of the
following methods.

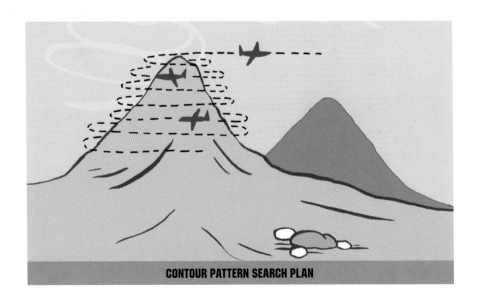

CONTOUR PATTERN SEARCH PLAN

LIGHT SIGNAL

Use a flashlight to send a Morse code SOS signal to an aircraft (see page 20). Alternatively, reflect sunlight with a hand mirror, the top of a survival tin, foil or glass to signal your position.

FLAG SIGNALS

Make a flag out of any brightly colored material, then send an SOS signal to an aircraft in the vicinity in the following way:

1 Swing the flag to the right (followed by a figure eight).

2 Repeat twice more to make the three dots for Morse code "S."

3 Swing flag to the left (followed by a figure eight).

4 Repeat twice more to make the three dashes for Morse code "O."

5 Repeat the sequence for "S" again.

▲ **MIRROR SIGNALS**
A mirror can be used to signal using reflected sunlight over a distance of many miles.

▶ **SAFE LANDING**
A British Commando Helicopter Force (CHF) Sea King Mk4 helicopter carries out training in the mountains surrounding Bardufoss, Norway.

38 Be weather-wise

Accurate prediction of the weather can have a decisive influence on your chances of survival in mountain regions, enabling you to make the best decisions about the choice of route or, if severe weather is imminent, whether to shelter.

UNDERSTANDING MOUNTAIN WEATHER

In a mountain range, weather conditions can alter radically in a very short space of time. It is therefore important to maintain constant awareness of the weather and to have a knowledge of early warning signs so that you can take appropriate action.

- Winds tend to be both stronger and more unpredictable than at low altitudes.
- Snowfalls are common above 8,000 feet (2,500 m) at any time of the year.
- If it rains in a mountainous area, you can expect flash flooding — especially in gullies — which can be dangerous. Heavy rains also increase the risk of rockfalls.
- Lightning can be especially violent during mountain storms.

SIGNS OF CHANGE

Knowing that the weather is about to change can dramatically influence your decision on whether to take shelter or attempt to make your way toward safety.

SIGNS OF WORSENING WEATHER

- Change in wind direction and increase in wind strength.
- Increase in cloud density and lowering of cloud cover.
- Rain becoming persistent.

- Drop in temperature.
- Increased humidity.

SIGNS OF IMPROVING WEATHER

- Steady wind of moderate strength.
- Rain decreases or stops altogether.
- Generally higher and sparser clouds.
- Rise in temperature.
- Reduced humidity.

STORM SENSE

If you are caught in a storm in the mountains, take the following potentially lifesaving actions:

- Steer clear of prominent features such as ridges that are more likely to be struck by lightning, and ensure you do not become a prominent feature yourself — by remaining on a rock wall, for example.
- Place wet ropes and metal objects at least 50 feet (15 m) away from you.
- Find shelter.
- Minimize the risk of being hit by falling rocks by leaning into a rock wall if you can't find shelter.
- Sit with your knees hugged close to your chest and your feet together.

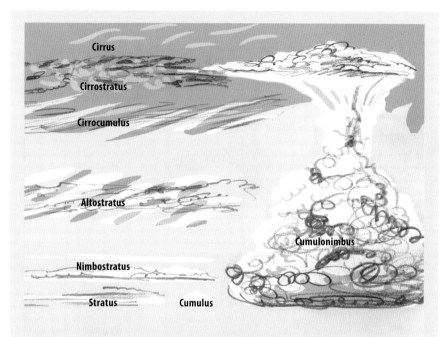

FORECASTING FROM CLOUDS

Clouds are one of the best indicators of likely weather conditions. They are normally grouped by altitude.

HIGH CLOUDS

Cirrus — Composed of ice particles. These appear as feathery bands and usually indicate fine weather. However, in a cold climate when accompanied by a steady wind they can indicate that a blizzard is on the way.

Cirrostratus — This veil of whitish/gray clouds can sometimes follow cirrus to indicate bad weather.

Cirrocumulus — Small fluffy balls of cloud arranged in groups. Indicates good weather.

MIDDLE CLOUDS

Altostratus — These form a grayish veil. If they become darker or thicker it usually indicates rain.

Cumulonimbus — With dark bases, they tend to be very tall, often with an anvil shape at the top. A clear indictor of bad weather: rain, snow or hail.

LOW CLOUDS

Cumulus — White fluffy clouds, often gathered together. Mostly indicate good weather.

Stratus — Low clouds formed in a gray layer. Indicate either rain or snow.

Nimbostratus — Form in low, dark blankets. Usually indicate rain or snow that may persist for several hours.

39 Making a mountain shelter

Making a mountain shelter is particularly challenging as you may be on a very steep incline. What's more, there may be few building materials available above the tree line. Rocks or snow can be used to make a good shelter if you cannot descend to lower ground.

ROCK SHELTER

In an area where there is no snow or natural shelters such as caves, you may need to create a stone shelter. On a sparse mountainside, there is unlikely to be anything to provide overhead cover but you may be able to create a roof with plastic sheeting or a poncho held down by stones. In this case, use a similar technique as that described for making a snow block shelter, using rocks instead of blocks of snow (see page 55).

SLEEPING ON A SLOPE

If you are sleeping on an incline, lie with your head up the slope and your feet down, to stop blood rushing to your head.

▼ ROCK WINDBREAK

In barren mountain environments, rock may be the only material available for building a shelter.

◀ USING SNOW DRIFTS
Here a snow cave is dug
directly into a snow drift.

MAKING A SNOW CAVE

Because of the exposed conditions on a
mountain, it is usually better to try to descend
to a more sheltered area if you are above
the snow line and you expect conditions to
deteriorate. However, if this is not possible,
a snow cave cut into a stable snow drift may
provide temporary shelter.

1 Cut directly into the drift and dig out a cavity
inside. It may be easier and more efficient to
cut the snow out in blocks.

2 Inside the cave, create a sleeping platform
that is higher than the entrance area.
Because warm air rises you should be
warmer in this area and the warm air will
not pass so easily out of the entrance.

3 Cut some breathing holes. Block the
entrance either with a rucksack or with a
block of snow to reduce the worst effects
of windchill, but take care not to seal off
all outside air.

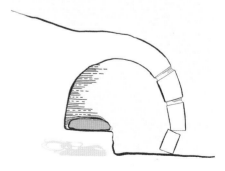

CUT BREATHING HOLES

40 Avoiding mountain sickness

Mountain (or altitude) sickness is a common and sometimes serious condition that can affect anyone at altitudes over 5,000 feet (1,500 m) above sea level, especially if the ascent is rapid. Susceptibility of individuals to the condition is variable.

SYMPTOMS

The symptoms of mountain sickness are not always felt immediately but are likely to become more extreme the higher you go. Severe mountain sickness — most likely at altitudes of over 10,000 feet (3,000 m) — may be disabling and can be dangerous, whereas you may be able to carry on almost as normal with milder symptoms that occur at around 5,000 feet (1,500 m).

MILD SYMPTOMS

- Headache.
- Tiredness.
- Nausea.

WHAT CAUSES MOUNTAIN SICKNESS?

Mountain sickness is caused by the drop in atmospheric pressure as height is gained. Lower pressure makes it difficult for the body to transfer oxygen from the lungs to red blood cells. As a result, there is a potentially dangerous increase in the level of carbon dioxide in the blood.

SEVERE SYMPTOMS

- Vomiting.
- Rapid and difficult breathing.
- Extreme weakness, listlessness.

ACTION

1 If possible, descend to a lower altitude for at least 24 hours to permit a more gradual acclimatization. Those who are severely affected may not be able to return to higher altitudes.

2 Rest and practice deep breathing.

3 Take a painkiller such as acetaminophen (if available) for the relief of any headache.

▲ DESCEND TO RECOVER

Marines carry one of their team to a lower altitude on a stretcher during a training exercise.

SIMPLE STRETCHER METHOD

A straightforward stretcher can be made if you have two poles at least 6 feet (1.8 m) long and a couple of sturdy jackets.

1 Pull up the zips and fasten any buttons on the two jackets, right up to the collar.
2 Pull the sleeves of the jackets inside out, so that they are hanging down inside the jackets.
3 Take each pole and pass it through the sleeve area of one of the jackets. Work the jacket along the poles until it is halfway along the poles.
4 Do the same with the other jacket.
5 You should now have an adequate and comfortable stretcher for carrying a sick or injured person.

MAKING A ROPE STRETCHER

If there are at least two people available and a third one is injured, creating a rope stretcher is an option. This may be the only option above the tree line where materials such as branches are not available.

1 Working from the middle of the rope, create 24 bights, each about 18 to 24 inches (45–60 cm) long.

2 Make a clove hitch (see page 186) over each bight.
3 Take the rest of the rope through the bights and outside the clove hitches.
4 Move the clove hitches toward the closed end of each bight and then tie off the ends of the rope with the clove hitches.
5 Make the stretcher more comfortable by lining it with a sleeping bag or clothing.

You can also slide two poles, if available, through the bights on each side to provide support. Then use two more poles to cross the head and foot of the stretcher, securing them with the ends of the rope.

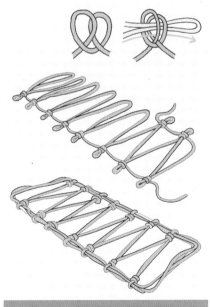

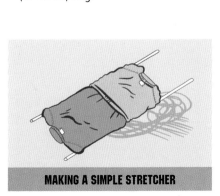

MAKING A SIMPLE STRETCHER

MAKING A ROPE STRETCHER

41 Be avalanche aware

Avalanches, common in mountainous regions, are a serious danger: they can sweep travelers away or bury them in deep snow drifts. Special Forces try to avoid avalanche-prone areas but they also learn survival techniques to use if they are caught in one.

KNOW THE DANGER SIGNS

Knowing that an avalanche could be imminent allows you time to take evasive or protective measures. There are various danger signals that Special Forces are trained to notice that alert them to the risk of avalanches, such as:

- An overhanging slab of wet snow with the wind behind it.
- New cracks appearing in the snow or signs of snow beginning to tumble.
- Hollow-sounding snow (meaning that it is not firmly bedded into the harder surface beneath).

- Signs that an avalanche has occurred previously in the area, including toppled trees and loose boulders and stones.
- Steep slopes. Avalanches usually occur on slopes that are between 30 and 45 degrees but can also happen on more gradual slopes.
- Shady mountainsides. Snow is more likely to shift dangerously on a slope that does not receive much sun, as it will not have melted and then formed ice.

◀ DEADLY DANGER
An avalanche is a constant and serious danger in mountain regions.

PRECAUTIONARY MEASURES

If you have to move across an area at risk of avalanche, take the following precautions:

- Maintain awareness of warning signs and of the weather conditions.
- Always stay roped to your companions.
- Traverse the slope as high as possible.
- Plan your route so that you pass close to any solid objects such as trees or large boulders that you can use as anchors if necessary.

HOW TO SURVIVE

If caught in an avalanche, use these Special Forces techniques to improve your chances of survival:

1 Remove any equipment, such as a backpack and skis.

2 Try to get to the side of the avalanche by using swimming movements.

3 While the avalanche is still moving, use swimming movements to help keep your head up and to stay near the surface.

4 If you are covered by snow, position your arms and hands in front of your face to form an air pocket and rock your head backward and forward to maintain a space in which to breathe. Keep as calm as possible to use as little oxygen as possible.

5 Do not try to fight the avalanche, but instead conserve your strength to get to the surface as soon as the avalanche has subsided.

42 Walking safely on mountains

On a mountain, you will always be on some degree of incline, and the ground may be slippery, consist of loose rock or shale or be covered in snow or ice. One of the keys to surviving on mountains is to know how to move safely in these conditions.

THE BASICS

There are various techniques for traversing different kinds of slopes and on different kinds of ground. The basic principles include the following guidance:

- Place your weight directly over your feet at all times.
- Avoid sharp movements or sudden stopping.
- Maintain a steady rhythm to avoid slipping and to conserve energy.
- Stop after an hour to adjust your equipment as necessary, including tightening your boot laces to provide maximum support.
- Drink enough water and eat enough food, if available, to replenish your energy.
- Move at a moderate pace to reduce the frequency of rest stops.
- Take off a layer of clothing or open your collar if you are beginning to sweat.
- Stop for rest at hourly intervals (approximately), depending on your fitness.

WHY WALK?

Walking in unfamiliar mountain territory is not something to be undertaken lightly, but in a survival situation, moving to a location where you are more likely to find help or where the conditions are less risky may be your best option.

▶ KEEP YOUR BALANCE
Special Forces like this U.S. Navy SEAL are trained to keep their body weight over their feet while walking on inclines.

TACKLING STEEP SLOPES

When climbing a steep slope you do not have the benefit of momentum as you would when walking on flat ground. Special Forces use the following technique, which enables you to both maintain forward propulsion and balance in the most energy-efficient way:

- Take a step forward and then pause for a moment, allowing the muscles in your forward leg to relax while placing your weight on the rear leg, which is locked.
- During the pause, breathe deeply and exhale strongly. This allows the necessary oxygen to refuel the muscles before you take the next step.

▲ PAUSE OFTEN

When moving on a steep slope, stay roped together and pause between steps.

AVOID JARRING

When walking downhill:
- Flex the knees of the forward leg to reduce jarring.
- Use an ice ax or stick to provide additional support.
- Dig in your heels to provide maximum grip.

43 Moving on snow-covered gradients

Movement on snow can be hard at the best of times, but it is especially so when you are on a steep gradient. Learning the techniques described here will make walking on snow-covered slopes easier and less risky, whether you are going up or down.

WALKING UPHILL ON SNOW

Conserve energy and reduce the risk of injury by following these guidelines:

- If you are moving directly uphill in snow, use an ice ax for support and kick steps in the snow before attempting to move upward.
- Try to avoid walking along the side of a hill as it places a great deal of lateral pressure on the legs and feet which can lead to injury, such as a twisted ankle, or loss of balance followed by a dangerous fall.
- If you are moving diagonally uphill, kick steps into the snow along your intended path and use an ice ax between your body and the hill to provide stability.

IF YOU FALL

Falling on a steep, snow-covered slope, leading to an uncontrolled downhill slide, can have potentially fatal consequences, especially if the slope ends in a precipice or other significant obstacle. Special Forces are trained to use a variety of methods to arrest a fall on a snow-covered slope.

BRAKING WITH AN AX

If you fall or slip on a snow-covered slope and have an ax on hand, you can use the following techniques:

ON SOFT SNOW

- Drive the shaft of the ax into the slope and your toes into the slope as well.
- Keep your hand near the base of the ax shaft.

◀ THE ESSENTIAL AX
An ice ax can be used as a support when walking uphill on snow.

> **SAFETY BREAK**
> If you fall in the snow an ice
> ax can stop your descent.

ON HARD SNOW

- Use the pick of the ax to grip the ice.
- Keep one hand on the head of the ax and keep the adze under the same shoulder.
- Place the other hand on the shaft.
- Roll over, pressing the blade of the ax into the snow by pushing down on it with the arm and shoulder.
- Keep your feet raised off the ground as you descend the slope.

BRAKING WITHOUT AN AX

If you fall or slip when you do not have an ax, use your limbs to slow and stop the fall.

- Roll onto your front and push up from the slope, straightening your arms.
- Keep the pressure on your toes to drive your toes and feet firmly into the slope. This will help to stop the fall.

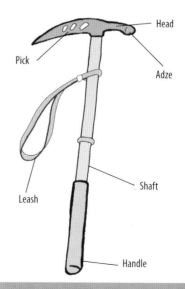

PARTS OF AN ICE AX

44 Descent by sliding

Glissade is a method of sliding down a snow-covered slope in a controlled way. You may need to use this technique to descend a slope quickly if, for example, you are aware that a storm is imminent and need to take shelter at a lower altitude.

WHEN NOT TO SLIDE

It is important to be aware of the times when it may be dangerous to glissade.

- Do not glissade when you cannot see the bottom of the slope that you want to descend. There may be a hidden precipice or other danger.
- Do not glissade when the snow cover is shallow — you may injure yourself.

STANDING GLISSADE

This is usually the fastest method of glissading. It provides the best view of the slope and potential obstacles. However, it also requires considerable skill to balance and you cannot use the ice ax for controlling the direction of your slide or for braking.

1 Stand in a well-balanced position, hold out your arms, and bend your knees slightly.

2 Allow the force of gravity and slippery ground to start your slide downhill.

3 Turn your body to change direction.

4 To stop, turn your feet sideways so that they dig into the snow.

STANDING GLISSADE

KNOW HOW TO BRAKE

The disadvantage of glissading is that it may turn into an uncontrolled slide if you are not careful. It can also damage clothing that is not made for tough conditions. A safe glissade requires competence at ice ax braking (pages 112–113).

CROUCHING GLISSADE

This method is a good choice when the ground is uneven. It allows a certain amount of control by the use of an ax for braking and direction.

1 Place one hand on top of your ice ax (or improvised alternative) and the other on the shaft and then crouch down.

2 Allow the force of gravity and slippery ground to start your slide downhill.

3 Drag the shaft end of the ax in the snow to balance. Press it into the snow to slow down or stop.

SITTING GLISSADE

Use this method when the snow or ice cover is smooth and there are no obstacles that could cause injury. A sitting glissade provides maximum control but your clothing is likely to be soaked by the snow.

1 Hold the ax in the same way as for the crouching glissade.

2 Sit on the snow and allow yourself to slide, increasing speed by raising your legs and leaning backward.

3 Slow or stop by pushing your heels into the snow and pushing down on the ax.

SITTING GLISSADE

45 Climbing safely without a rope

If you are stranded in a mountainous area there is a possibility that you may not have a rope. In a survival situation, basic knowledge of how to move on a rock face could help to avert danger.

WHY CLIMB?

Special Forces are trained to climb (and descend) comparatively short rock faces without rope in order to surprise an enemy or to avoid detection. In a survival situation, you may need to undertake a ropeless climb or descent to avoid danger or to reach a place from which rescue is possible. Never attempt a long ascent or descent without a rope.

On these pages there are some key holds for hands and feet as well as some climbing moves for tackling a steep rock face.

KNOW THE PRINCIPLES

Before you attempt a climb, take a good look at the rock face and try to spot the best hand and foot holds so that you can work out the route that you might follow. This will help when you are on the rock face and it is more difficult to see ahead. Remember these climbing guidelines:

- Use four limbs plus your torso in as smooth and coordinated a way as possible.
- Move only one of these parts at a time.
- Use the legs to support your weight, and your arms and hands.

HOW TO MOVE ON A ROCK FACE

The key to moving safely is to maintain your balance at all times by not moving your foot too far at one time.

SHIFT WEIGHT TO RIGHT FOOT

MOVE LEFT FOOT

WEIGHT OVER BOTH FEET

LIFT BODY WITH BOTH LEGS

1 Shift your weight onto one foot so that the other can be moved to a new position.

2 Put your weight back onto both feet and move your body into the new position.

3 Move one hand to a new position that is within an area between your head and your waist, making sure that you maintain your balance with both legs and the other hand.

4 Then move the other hand according to the same principles.

SAFE HANDHOLDS

The most common form of handhold is the pull hold, where the hands and fingers cling to a ledge or protrusion in the rock. Do not be tempted to use this hold for bearing weight. Other ways of clinging to the rock with the hands include the down cling, side cling and opposing side clings.

SAFE FOOTHOLDS

Placing your feet correctly enables you to use your legs to maximum effect, lifting your body while minimizing the risk of slipping. This also helps to minimize tiring of the arms. To keep a good foothold:

- Maximize the contact of the foot/boot with the rock.
- Maintain maximum friction by keeping as much of the boot sole as possible on the rock and keeping the heels as low as possible (see below).

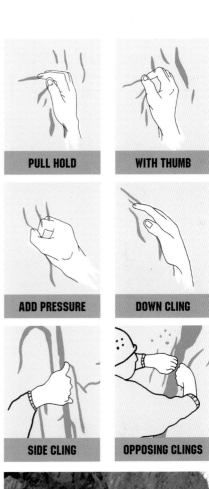

PULL HOLD

WITH THUMB

ADD PRESSURE

DOWN CLING

SIDE CLING

OPPOSING CLINGS

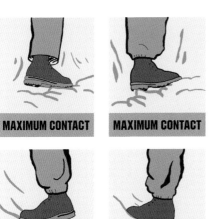

MAXIMUM CONTACT

MAXIMUM CONTACT

MINIMUM CONTACT

MINIMUM CONTACT

46 How to safely anchor a rope

When surviving in the mountains, having ropes available can be a huge bonus, enabling you to move more safely down steep slopes. A secure anchor is a prerequisite for any form of rope-assisted descent.

A SECURE SUPPORT

Before undertaking any descent in the mountains that relies on rope support, Special Forces pay particular attention to ensuring that the anchor point is not only strong enough to bear the load of the person or people on the rope (as well as any equipment), but is also able to withstand shock loads that might occur from a fall. The rope and knots that are attached to the anchor need to be similarly strong and securely attached.

SUITABLE ANCHOR POINTS

The anchor may be fixed to any of the following, or spread between more than one solidly fixed object:

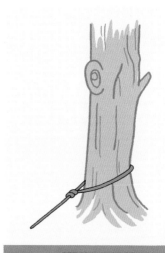

STRONG TREE TRUNK

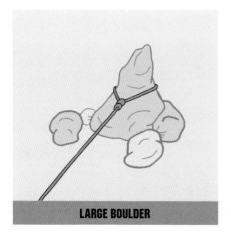

LARGE BOULDER

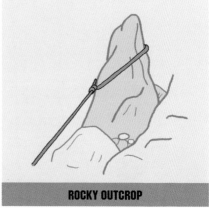

ROCKY OUTCROP

FIXING THE ANCHOR

The bowline (see page 185) is a strong knot that can be used to secure an anchor for a rope-assisted descent in the mountains. Tie the bowline around an anchor point as follows:

1 Pass the working end of the rope around the anchor point from right to left (facing the anchor). Take the standing part of the rope and form an overhand loop toward the anchor.

2 Pass your hand through the loop and pull through a bight of rope.

3 Pass the working end of the rope through the bight and bring it back on itself. Tighten the knot. Take the tail of the bight and form an overhand knot (see page 183). There should be at least 4 inches (10 cm) of rope free beyond the safety knot.

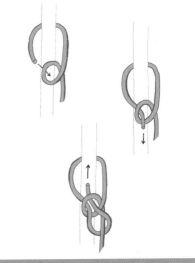

SECURE AN ANCHOR

◀ **CLIMBING DOWN SAFELY**
When using ropes as part of a descent, it is often preferable to set up a main rope plus a safety rope.

47 Getting down quickly

Knowing how to get down a moderately steep slope safely and quickly can be vital to survival — for example, if bad weather suddenly threatens. In this situation, a technique known as hasty rappeling can be done with minimal equipment.

MAKING THE DESCENT

Use this technique to get down a slope quickly. First, make sure that you are descending to a safer area. You'll need a rope that is long enough to reach the bottom of the descent after the anchor has been fixed. This method should not be used for making descents that are vertical or very steep.

1 Having fixed your anchor securely (see page 119), position yourself side-on to the anchor and place the rope horizontally across your back, maintaining a firm grip.

2 Walk down the slope sideways, allowing the rope to pass through your hands.

3 To slow or stop, bring your brake hand (the hand controlling the lower end of the rope) in front of your body — this will lock the rope — and turn to face the anchor.

SIDEWAYS STANCE

TAKE SMALL STEPS

BRAKE POSITION

DESCENDING WITHOUT A ROPE

If you are walking down a mountain without the aid of a rope, the best technique to use varies according to the gradient you are on.

STEEP GRADIENT

For steeper gradients, face the rock as you descend. Balance your weight on your legs and keep your legs bent and arms straight.

MODERATE GRADIENT

If the gradient is less steep, move sideways but always maintain three points of contact — either two hands and one foot or one hand and two feet.

SHALLOW GRADIENT

On a shallow gradient, you can descend face out, which allows you to see the ground ahead and enables you to find secure footholds where you can jam your boot to stop yourself sliding on the rock or, if you are moving across a grassy or snow-covered slope, to kick the footholds you need.

▶ FACE OUT
A Marine practices a "face out" descent of a rocky slope using the hasty rappel technique.

48 How to travel over a glacier

Glaciers are large bodies of permanent snow that over time have hardened into ice. They can provide a clear route down a mountain to safety, but can also present risks for unwary travelers.

KEEPING CLEAR OF CREVASSES

A key risk for anyone who needs to move on a glacier is that of crevasses. These are cracks caused by the stresses of glacier movement. To minimize the chance of falling into a hidden crevasse, it is important to know how to spot the signs of their presence and how to travel in maximum safety.

JUMPING A CREVASSE

If you attempt to jump a crevasse, cross it at right angles and give yourself enough momentum to fall forward after the jump, ensuring that you also dig in on the far side with your ice ax for extra security.

▶ SAFETY ROPE
A group traverses a dangerous glacial ridge, roped together to provide crucial support in the event of a fall.

123

WALKING SAFELY

Follow these guidelines to keep you safe on a glacier:

- Rope yourselves together if two or more people are moving across a glacier. Allow 40 to 45 feet (12–14 m) between each person. The distance between walkers reduces the chance of two people falling into a crevasse together.
- Maintain these distances even when stopping for a break. The combined weight of the whole party might be enough to collapse a crevasse bridge.
- Have little or no slack in the rope in order to minimize the distance that any individual might fall.
- Do not carry coils of rope in the hand but instead pass the coiled rope around the body diagonally, over one shoulder and under the other. This way, if a climber ahead falls into a crevasse, the rope will not be ripped out of your hands; the shock on the rope can be absorbed by your body. The holding climber can then sit down and take up a bracing position to support the fallen climber.
- Provide maximum anchor force if a member of the party falls into a crevasse. Every other member of the group should immediately lean back on their haunches and dig their heels into the snow.

AVOIDING AREAS OF WEAKNESS

When crossing a glacier, be alert for signs of weak snow, such as hollows. Additional safety tips include:

- Carry a stick or ice ax to probe the snow in suspicious areas. If the snow collapses, move back and cautiously try another route.
- When moving, look out for places where the glacier becomes steeper or bends — a crevasse is most likely to form in these areas.
- If possible, choose to cross a glacier in the early morning while it is still cold and the ice has not begun to melt.

49 Climbing out of a crevasse

Falling into a crevasse is perhaps the greatest risk you face when crossing a glacier. If this happens, a calm and methodical response can be lifesaving.

PRUSIK SELF-RESCUE

If a companion or a team at the top are unable to rescue one of the party who has fallen into a crevasse, but they are still securely roped to the team, the person will need to perform a self-rescue. The most usual method of self-rescue is by use of the Prusik system. If possible, when undertaking a journey across a glacier, prepare for this eventuality by carrying at least two Prusik loops and climbing slings (loops of webbing), which are essential to the technique. Extend one Prusik loop with a sling so that you can put your foot in it. The loops should be clipped out of the way on the back of your harness until they are needed. It is a good idea to have a couple of spares.

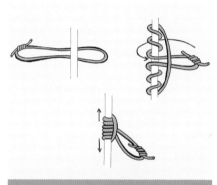

TYING A PRUSIK LOOP

MAKING A PRUSIK LOOP

Make a Prusik loop out of ¼-inch (5 mm) cord. Tie a fisherman's knot in 4 to 5 feet (1–1.5 m) of cord (see page 184).

ATTACHING A PRUSIK LOOP

The key to the Prusik self-rescue is the technique for attaching the loops to the rope you are climbing up (the climbing rope).

1 Wrap the loop around the climbing rope.
2 Pass the join back through the loop.
3 Make more wraps for added security.
4 Tighten the knot around the climbing rope. (See above.)

◀ **THE RIGHT STUFF**
Mountain-climbing gear and ice axes are essential and lifesaving tools.

USING PRUSIK LOOPS FOR SELF-RESCUE

This is the method practiced to use two Prusik loops to escape a crevasse:

1 Let go of your backpack and let it hang below you on a drop cord.

2 Attach the shorter Prusik loop to the climbing rope above the chest harness attachment.

3 Straighten out the wraps so that they are not crossed to improve the grip.

4 Pull straight down to check that the loop is holding and, if not, make an extra wrap or adjust the existing wraps.

5 Attach the extended loop beneath the first loop but above the chest harness attachment point.

6 Check that the loop is holding, and insert your foot into the loop extension.

7 Slide the first loop up the rope as far as you can and hang from it while you move the Prusik supporting the foot loop up the rope.

8 Stand on the foot loop extension and move the first loop up the rope again.

9 Repeat the procedure until you reach the lip of the crevasse. Then grasp the climbing rope with both hands and pull yourself onto level ground.

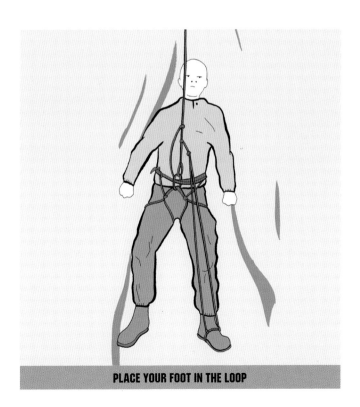

PLACE YOUR FOOT IN THE LOOP

50 Dealing with fractures, breaks and strains

Broken or sprained limbs as a result of falls or slips while walking on rocky, steep terrain are very common. Knowledge of how to treat such injuries is vital, as obtaining immediate outside help may be difficult because of the steep gradient and unpredictable conditions.

HOW TO IDENTIFY A FRACTURE

Signs of a fracture (broken bone) include:
- Swelling, bruising and distortion of a limb beneath the skin.
- A grating sound if the limb is moved.
- Pain when the limb is moved.
- Bone visible through an open wound.

STABILIZING AN INJURY

In a survival situation it is unlikely that you will be able to reset a fractured bone. The best thing you can do is to immobilize it with a splint to reduce pain and prevent further damage. If the broken bone has broken the skin, place a clean dressing over it as soon as possible.

ARM INJURIES

Suspected fractures of the shoulder, arm or elbow should be immobilized with bandaging and a sling.
- A triangular bandage or folded square of fabric can easily be used as a sling.
- Crepe bandages are ideal for binding. Special Forces usually carry rolled crepe bandages in their medical kit.
- Never apply bandages too tightly as this will interfere with circulation. Do not extend bandages by tying knots in them; instead, bind a separate layer of a new bandage over the previous one.

◀ HELP FIRST
In survival situations in the mountains, first aid is the top priority, then shelter, water and food.

TREATING A STRAIN

A strain, sometimes referred to as a pull, usually results from tearing of the muscle tendon, and can easily occur as a result of a fall or other accident. There may be pain, stiffness and swelling. Rest, Ice, Compression and Elevation (RICE) is the standard treatment for soft-tissue injuries of this kind.

1 Apply cold to the injured area as soon as possible. Ice or snow wrapped in a cloth is ideal.

2 Compress the injured part by wrapping it firmly — but not so tightly as to restrict blood flow — with a bandage or improvised alternative.

3 Rest with the injured part elevated for as long as possible.

HOW TO MAKE AND APPLY A SPLINT

The aim of applying a splint is to keep the bones from shifting out of alignment by preventing movement of the broken limb. If a leg is fractured and you do not have a suitable splint, bind the broken leg to the sound one to provide support.

1 Use any solid materials that are available, including lengths of wood or sticks.

2 Place some padding material, such as spare clothing, between the splint and the limb.

3 Tie bandages or pieces of cloth around the limb, the padding and the splint to hold it all firmly in place — but do not tie the bandages too tight as this will restrict blood flow.

LOWER-LEG FRACTURE

Make a splint that extends from above the knee to the below the foot. If there are no splints available, bind both legs together.

HIP OR UPPER-LEG FRACTURE

Place a splint on the inside leg and a longer one extending from the armpit to the ankle. If no splints are available, place a blanket between the legs and bind them together.

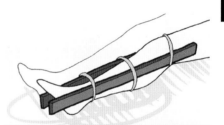

LOWER-LEG FRACTURE

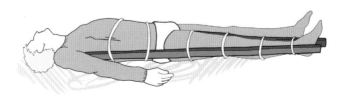

HIP OR UPPER-LEG FRACTURE

4 JUNGLE SURVIVAL

The jungle is one of the environments most closely associated with Special Forces, and one in which they have unmatched expertise. Jungle warfare schools operate in various parts of the world to teach the vital skills required for military operations and survival in this terrain.

TESTING CONDITIONS IN THE TROPICS

The heat and humidity of a tropical jungle present specific survival challenges. Although the environment is teeming with life, for those unused to the conditions, the jungle can be fraught with danger from the risk of disease to the difficulty of finding your way to safety.

JUNGLE SURVIVAL

130

Tropical areas cover a large landmass on either side of the equator between the Tropic of Cancer in the north, and the Tropic of Capricorn in the south. The principal regions around the world are the tropical rainforests of Central and South America, Africa, India and Southeast Asia. These areas are characterized by high humidity and lush vegetation. There

are often bouts of heavy rainfall, which start and stop suddenly. Because these regions are close to the equator, the sun also rises and sets relatively quickly.

The word jungle generally conjures up images of dense forest and undergrowth, and this is true of primary jungle. However, environments known as "secondary jungle"

◀ DENSE AND HUMID
A U.S. Marine patrols through the jungle as part of a cover and concealment class during jungle warfare training. The density of the jungle is a problem in survival situations as it is difficult to find crash sites.

◀ **JUNGLE NAVIGATION**
A U.S. Marine navigates through the jungle with a compass during jungle warfare training.

▼ **FIND SHELTER**
A basic shelter and campfire in the jungle regions of Thailand.

include areas that have been cleared of large trees, but which nevertheless have a thick undergrowth. Jungle areas also include seasonal and monsoon forests, scrub, savanna and thorn, saltwater and freshwater swamps.

OBSCURED BY TREES

In the primary rainforest there are several layers of tree canopy. The density of the canopy means that rescue from a jungle area of this type is unlikely as no aircraft is liable to see you on the ground and signals will probably be ineffective unless you can find a clearing. Your best chance of escape may be to walk to safety, but in the jungle it is often hard to see the sky clearly. It therefore may not be possible to navigate using either the sun or stars, and it is often very difficult to get a

sense of direction. Jungle travel and navigation is highly specialized — Special Forces always carry a compass in jungle regions and carry out intensive training in similar environments in order to master this skill.

SPECIAL JUNGLE EQUIPMENT

Special Forces operating in the jungle are careful to equip themselves with the right tools and supplies to enable them to survive the challenging conditions. A key tool for any team is the machete, also known as a parang in Southeast Asia. This is a large knife with a blade that is usually between 12 and 18 inches (30–45 cm) long and up to $^1/_{10}$ inch (3 mm) thick. The machete is very useful in jungle environments for hacking a path through forest undergrowth and for cutting materials to build shelters or rafts. If Special Forces are not carrying a standard issue compass, they often have a button compass in their emergency survival kit. Although apparently basic, the button compass is oil-filled and pressurized to provide accurate readings. It can easily be sewn into the lining of clothing, and can also be fitted to a watch strap.

WATER, FOOD AND SHELTER

Perhaps surprisingly in an environment in which rainfall may be high, finding water is an important consideration in the jungle as high temperatures mean that you will sweat copiously. While there is plenty of water in the jungle — in waterways and pools — the key is knowing where to find sources that are safe to drink.

Food is also plentiful but you need to take care to choose the right plants. The jungle is crawling with animal life and, although some creatures are edible, many species can be dangerous or may at least cause discomfort. Some of the more dangerous insects and

◀ **ESSENTIAL KIT**
Special Forces are trained to use machetes for the construction of shelters during jungle warfare training.

◀ **FIND SAFE WATER**

Forces are trained to filter water through layers of camouflage material to help make it safe to drink.

▼ **DEADLY INSECTS**

Some jungle wildlife look deadly and even small insects can be a serious hazard.

wildlife include marching ants, mosquitoes and snakes. A strong insect repellent to deter attack and antiseptic to treat bites and scratches are key items in any jungle survival pack — if you don't have specialist supplies, you will need to prevent bites by covering bare skin. Another related risk in jungle conditions is that of infection developing in an open wound. Scrupulous attention to the protection of scratches and bites is essential.

It is also extremely important to build adequate shelters and to create safe places to sleep in the jungle. Although the climate is warm and you are not subjected to extremes of temperature (as in the desert or polar regions), it is important to keep yourself off the forest floor, to avoid attack by insects and reptiles during the night. A hammock is a highly useful lightweight solution to this problem. The hammock can be strung between two trees and a poncho or plastic sheeting can be positioned above it to keep rain off.

Some Special Forces may be issued with a variety of NATO standard hammocks, the lightest of which may weigh only 11 ounces (310 g). A mosquito net is a key part of any jungle survival kit.

JOURNEY TO SAFETY

Because of the difficulties of summoning assistance in the jungle and the necessity in most cases of finding your own way to safety, this chapter includes descriptions of key Special Forces techniques for traveling through jungle and across rivers.

The jungle is a hugely demanding environment, but by using specially devised survival techniques, based on extensive experience in such environments, you can boost your chances of overcoming the dangers of the jungle and reduce your risk of illness or injury.

51 Getting down from the jungle canopy

Special Forces may be parachuted into a tropical area, and land accidentally in the jungle canopy. Others may be caught in the canopy by a forced parachute landing after an aircraft system failure. Knowing how to get down is a key survival skill.

THE CHALLENGE

Simply climbing down a tree from the high jungle canopy is not usually a realistic possibility as branches can be widely spaced and may come to an end too high above the ground for a jump to be made safely. The best and safest option is to use a rope, if available. This method assumes you are wearing a harness and have a carabiner available. Before you lower yourself, drop any equipment to the ground first. Special Forces practice this type of descent extensively before making a drop into a jungle area.

ROPE-ASSISTED DESCENT

1 First, attach the rope to a branch large enough to support your weight. Use a secure knot such as a figure eight loop or a bowline (see page 185).

2 Wind loops of the rope around your carabiner — opposite the opening mechanism — five times.

3 Clip the carabiner into the two short slings of your harness. Feed the length of rope attached to the tree into the carabiner from the top, while the rope that hangs below emerges from the bottom. Ensure that the carabiner is locked.

4 Check there is no slack between the carabiner and the tree and make sure that you are holding the end of the rope hanging below you securely in one hand.

5 Test the system by pulling in some rope until the safety sling goes slack and hold yourself on the rope. If you think there is not sufficient friction for efficient braking, make some more wraps around the carabiner. Conversely, there may be too much friction in the carabiner, in which case take out a couple of wraps.

6 When you are ready to descend and have firm hold of the rope, unclip from the safety sling.

7 Keeping firm hold of the rope, lower yourself gently. Use the rope hanging below you to control the rate of descent. Take care not to descend too fast. Once you reach the ground, simply unclip yourself from the rope.

▶ PREPARED IN ADVANCE

A Marine rappels to the ground during jungle warfare training in Colombia. Special Forces receive extensive training about the importance of forward planning to make self-rescue as easy as possible in an emergency.

52 Getting rescued

Signaling for rescue in the jungle can be a major problem due to the density of the tree canopy. Signal fires may be your best option if your position isn't obvious from the air.

MAXIMIZE YOUR VISIBILITY

If your survival predicament results from an aircraft crash, the best signal is the crash site itself. An aircraft will either leave a hole in the canopy or if it is resting on top of it, will be visible from the air. If you have jumped from an aircraft by parachute, maximize your visibility by leaving the parachute open at the site. Additional ways to boost your chances of being seen include the following:

- Place brightly colored material in the branches that will be visible from above, or use a stick to push a signal flag through the canopy, if you can safely climb high enough.
- Try to find a clearing in the jungle where you can display a contrasting color, using either fabric or smoke.

- Set up a marker on an island in a river, or prepare a fire on a platform or raft, which can be lit without danger of causing a forest fire (see opposite).

HOW TO CREATE A SIGNAL FIRE

Due to the thickness of the jungle canopy and the risk of starting an uncontrolled fire, you need to find adequate space to set up a fire that is large enough to be seen but will not cause a conflagration. Here are some tips:

- Set your fire in a jungle clearing. If necessary, build an earth rampart around the fire to prevent it from spreading.
- Burn a single tree in a clearing by adding bits of tinder between the boughs and any other flammable material that will ignite the foliage.

◀ GETTING SEEN
The thick jungle foliage makes spotting survivors difficult.

- Create a cone-shaped signaling platform (see right).
- Burn green leaves, green grass and ferns to create the maximum amount of easily visible white smoke.
- Have your signal fires ready to light at short notice; noise does not travel far in the jungle so you are not likely to hear an aircraft approach until it is very near.

MAKING A SIGNALING PLATFORM

A signaling platform can be used to drape brightly colored material over during the day and be filled with dry tinder and foliage to light if an aircraft should appear at night. Quickly pull off the material before lighting the fire.

1 Lash three long sticks together at one end to form a tripod.

2 Lash three shorter sticks together to form a triangular frame. Lash the frame between the splayed legs of the tripod.

3 Lay further sticks across the triangular frame and bind in place.

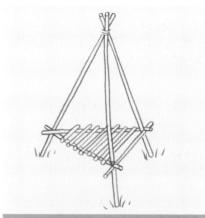

PUSH THE TRIPOD INTO THE GROUND

MAKING FIRE RAFTS

A safe way of lighting a signal fire in the jungle is to create fire rafts to float on a river. Attach the rafts in a triangular shape to create an open visible space for the fire as well as keeping the flames away from surrounding vegetation. Attach the rafts to the bank to prevent them from floating downstream.

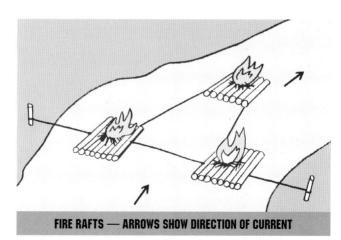

FIRE RAFTS — ARROWS SHOW DIRECTION OF CURRENT

53 Jungle clothing and equipment

Due to the unusual humidity in the jungle, clothing is often almost continuously damp – from sweating or heavy rainfall. To stay healthy, you need to try to wear clothing that protects you, but is also light enough to dry quickly.

COMBATING DAMP

Special Forces operating in jungle conditions wear clothing that is made from a light fabric, which is likely to dry more quickly than that used for a conventional temperate uniform. Clothing made from modern, high tech fabrics can often be more comfortable than items made from conventional fabrics such as cotton.

JUNGLE FOOTWEAR

When choosing footwear for the jungle, features that Special Forces look out for include the following:

- Rot-proof canvas uppers.
- A metal plate in the sole to prevent sharp objects penetrating it and injuring the foot.
- Drainage holes in the uppers to allow water and moisture to escape. If your boots do not have these holes built in, during a survival situation you can cut them yourself.

◀ LIGHTWEIGHT CLOTHES

Jungle clothing should cover as much skin as possible to avoid insects — it also needs to dry quickly.

◀ NET GAIN
In the jungle a mosquito net is essential. A hammock also keeps you safe.

PROTECT YOUR HEAD

In the jungle, a hat with a brim is an invaluable protection from insects. It helps to prevent insects or debris from falling inside your shirt, and soaks up sweat.

KEEPING COVERED AND CLEAN

Here are some tips used by Special Forces for getting the maximum protection from clothing in the jungle:

- Wear sleeves rolled down and trousers tucked into boots to provide maximum protection from biting insects and other forms of bites and stings.
- Wash clothing in clean water, if there is enough available, every two days or so. Use some soap or other detergent, if available. Due to the humid conditions, clothing may begin to rot if not regularly cleaned.

EQUIPPED FOR SLEEP

In order to maintain maximum alertness on operations, Special Forces recognize that it is important not to let damp clothing or insect attacks interfere with a good night's sleep. They often take two sets of uniforms on jungle operations. You can learn from their experience to maximize your survival chances.

- Keep one set of clothing dry for sleeping in.
- Hang out clothes worn during the day to air and dry as much as possible during the night.
- Change back into your day clothing in the morning, even if it is still cold and clammy — it will soon warm up in the heat and activity of the day.
- Use a mosquito net or a hammock. The advantage of a hammock is that it enables you to sleep off the jungle floor, is relatively easy to set up between two trees and is also light to carry.

54 Avoiding insect bites

Bites and stings from insects are among the most common and also potentially serious hazards of the jungle environment. Bites may be dangerous in themselves, or may lead to infection. Special Forces learn to recognize, avoid and deal with biting insects.

KNOW YOUR ENEMY

MOSQUITOES
One of the most potentially dangerous insects in the jungle is the malaria-carrying female anopheles mosquito. Depending on the region, this mosquito can also carry other diseases, such as yellow fever and dengue fever. The female mosquito normally feeds between dusk and dawn, which is why it is vital to use a mosquito net when sleeping if at all possible.

ANTS
A good reason for not sleeping on the ground is the potential presence of army or legionary ants. These ants move in large columns that can be up to 30 feet (10 m) wide and will devour anything in their path. Another species, the fire ant, can inflict a painful bite.

SPIDERS
There are more spiders in the tropics than anywhere else. Spiders such as the black widow have a life-threatening bite but

DEALING WITH STINGS AND BITES

Bites and stings from insects and other creatures can cause a dangerous allergic reaction, so treat them as soon as possible.

INSECT BITES
Wash the area (with soap as well as water, if possible) or apply a cold compress.

BEE STINGS
Stroke the area with the side of a needle to remove the sting or brush it out with your fingernails to avoid bursting the venom sac. Apply a cold compress to the wound.

TICK BITES
To remove a tick, wear gloves if possible and use tweezers (or clean fingernails) to pull the tick directly out of your skin, grabbing it as near to the skin as possible. Clean the area of the bite with iodine or other antiseptic.

MOSQUITO

ANT

TICKS

SPIDER

BEE

WASP

HORNET

most spiders are less dangerous. The famed tarantula of the South American jungle has a more fearsome size than bite. As a rule, it is a good idea to avoid all spiders if possible and especially those that are dark-colored or that have red, white or yellow spots. Spider bites should be treated urgently in the same way as snake bites (see page 159).

BEES, WASPS AND HORNETS

These insects can be very aggressive and inflict painful stings if they are disturbed — most commonly if you accidentally damage their nest. Nests are often attached to tree trunks. The options are either to sit still until the swarm has settled down and then move away slowly; to run through thick vegetation, which may provide protection or to immerse yourself in water, assuming there is a river nearby. Do not forget to check whether there is something more dangerous lurking in the water.

HOW TO DEAL WITH LEECHES

Although leeches — often present in jungle ponds and rivers — are not particularly dangerous, they do suck human blood and are unpleasant. More importantly, the bites can lead to infection.

WHAT TO DO

Pulling a leech off may not work, as the jaws can remain embedded in the flesh and increase the risk of infection. Apply salt, a glowing ember or the end of a cigarette to the body of the leech to make it release its grip.

LEECH IN THE JUNGLE

55 Getting water from plants

In the hot and humid conditions of the jungle, the human body cannot survive without water for longer than three days. If rivers, streams or collected rainwater are not available, an important alternative source of water is water-storing plants.

WATER FROM BAMBOO

Bamboo is a form of grass with a woody stem. Water is carried inside the stem to the rest of the plant.

- Using a sharp knife, cut into the stem and then, after checking that the liquid is clean and clear, drain it either directly into your mouth or into a container.
- An alternative method is to cut the bamboo stem near the top and then bend it back toward the ground and secure it in position with a rope. Place a container under the cut end of the bent stem so that it catches the liquid as it drips from the stem.

WATER FROM BANANAS

The edible banana fruit is itself 75 percent water, providing a double benefit of fluid and nutrition. There is also plenty of water within the "trunk" (formed from the base sections of the leaves) that is easily accessible.

1 Using a saw, cut through the trunk of the banana tree about 12 inches (30 cm) above the ground.
2 Use a knife to hollow out the stump to create a bowl. The bowl area should begin to fill with water. The water may have a bitter taste, but it is safe to drink.

POISONOUS PLANTS

When drinking water from plants, be aware that some are poisonous and any water that collects on the leaves or inside the plant may be tainted. Apart from a few well-known exceptions such as coconuts, the general rule is: do not drink from any plant that produces milky or colored sap. See also Poisonous Plants, pages 144–145.

LET THE BASIN FILL WITH WATER

WATER FROM VINES

The jungle vine or liana is a thick woody climbing plant that has a hollow center to carry water to the rest of the plant.

1 Cut into the vine as high as possible and then cut through the whole stem near to the ground.

2 Hold the cut vine over a container to collect the liquid. If the water is not clear or if it causes any irritation in the mouth when you dab a small amount on your lips, try a different kind of vine.

3 Once you know the vine is safe, you can drink directly from the vine: hold it above your head and allow the water to drip into your mouth.

CUT THE VINE AS HIGH AS POSSIBLE

◄ JUNGLE HYDRATION

Drinking from a cut vine during a U.S. Navy Jungle Survival School exercise.

56 Poisonous plants to avoid

Plants are a valuable food source in the jungle, but some of them are poisonous and can kill you. It is therefore a great advantage to know the dangerous plants in the area you are visiting.

KNOW THE DANGERS

Knowledge of poisonous plants that you should avoid eating could save your life in the jungle. Some poisonous plants give clear warning of the danger. Avoid any plant with white or yellow berries, all of which are poisonous, and also take great care to identify correctly any plants with red berries, half of which are poisonous. Avoid any plant with milky sap and any plant that irritates the skin when touched, and steer clear of plants with umbrella-shaped flowers. These are some examples of the most common poisonous jungle plants to reject as a possible food source.

IF YOU SUSPECT POISONING

- If you have eaten a poisonous plant, induce vomiting as soon as possible by putting your clean fingers down the back of your throat.
- If you have been in contact with a poisonous plant, wash the skin with soap and water and remove any affected clothing.

CASTOR BEAN

Found in tropical Africa, the seeds of this plant contain the poison ricin.

LANTANA

Belonging to the verbena family, this plant is found in both tropical and temperate areas — it has poisonous blackberry-shaped fruit.

MANCHINEEL

Also called poison guava, this plant is found in the southern United States, Central America and northern South America. It is known for its poisonous yellow-to-reddish, sweet-scented, apple-like fruits, which often grow in pairs. Smoke from a fire made from its wood can inflame the eyes and contact with latex from its leaves and bark may cause skin irritation.

WATER HEMLOCK

This parsley-like plant is found worldwide near water. All parts are highly poisonous — a small mouthful can kill an adult. Avoid handling the plant as even a small amount of its juice transferred to the mouth can cause serious illness. It has purple-streaked stems and small white flowers.

PANGI

Found in Southeast Asia. It has heart-shaped green leaves, green flowers, and produces brown nuts. All parts of this plant, including its pear-shaped fruit, are poisonous.

STRYCHNINE TREE

Found in Southeast Asia and Australia, this is an evergreen tree with greenish flowers and orange to red berries. If this plant is eaten, the resulting severe cramps and spasms may cause asphyxiation and death.

57 Plants you can eat

There are many edible plants in the jungle that can provide you with food even if you do not find any meat. Special Forces are trained to recognize edible plants in their area of operations. If in doubt, test for edibility as explained on page 149.

TOP TEN EDIBLE PLANTS

Although there are some plants in jungle areas that are easily recognizable as edible, such as banana or coconut, it is best to be cautious about other plants unless you can positively identify them as edible or have carried out an edibility test (see page 149). Do not assume it is safe to eat a plant simply because you have seen animals eat it — it may be poisonous for humans. The plants shown on these pages and on page 148 are just a few of the plants found in jungle areas that you can eat.

BETTER SAFE

Always err on the side of caution when considering any plant as a potential food source. It is better to go hungry than to deal with the effects of poisoning.

1 BAMBOO

The young shoots are edible, but take care to remove the poisonous black hairs found on the edge of the leaves. Bamboo seeds can be boiled and eaten.

2 BANANA

This is a highly nutritious fruit. It is less well known that you can also eat banana flowers, roots and leaf sheaths if you boil them first.

3 MANIOC

Also known as cassava, this plant has edible tuber-like roots. Take care to avoid the bitter version of the plant, which contains poisonous acids. If you cannot distinguish between the sweet and bitter varieties, or don't know how to prepare manioc safely, it's best to avoid it all together.

4 NIPA PALM

This plant is found in estuaries in Southeast Asia. It has a fern-like appearance and produces an edible fruit. The flower stalks and seeds can also be a source of both food and water.

5 COCONUT

This is a valuable source of both liquid (coconut milk) and food. The coconut sheath from which the leaves protrude can also be boiled and eaten.

6 RATTAN PALM

This tree can be found in the tropical rainforests of Africa, Asia and Australia. It produces a white flower. The tips of the stem and the palm heart can be eaten, either roasted or raw.

7 SUGAR PALM

This large-leaved plant is widespread in the tropics. The flower seeds can be boiled and eaten.

8 WILD YAM

This ground-creeping plant is found throughout the tropics. The root can be boiled and eaten.

9 TARO

This common tropical plant has large, heart-shaped leaves. You can eat the roots, young leaves and stalks either boiled or roasted.

10 WATER LILY

Found in both temperate and subtropical regions, it has edible seeds and roots, which can be either boiled or roasted before eating.

TESTING FOR EDIBILITY

When Special Forces want to find out whether a plant is poisonous or safe to eat, they conduct an edibility test. It is not worth carrying out the test unless you think that the plant may be a good source of nutrition, as it will take up most of a day.

DOS AND DON'TS FOR EDIBILITY

- **Do** test different parts of the plant separately; some parts may be edible and some not.
- **Don't** try to test more than one plant at a time.
- **Don't** test any fungi or any plants with a milky sap, except for those that you know to be safe, such as coconut. Others are highly likely to be poisonous.
- **Don't** eat anything else for eight hours before the test.

HOW TO DO THE TEST

If you notice an adverse reaction at any stage, reject the plant as a possible food source.

1 Test the plant for any skin reactions by crushing a section of the plant and rubbing the sap on a sensitive part of your skin, such as the inside of your wrist or elbow. Wait about 15 minutes for any reaction.

2 Take a small amount of the plant and put it on the outer surface of your lip. Wait three minutes for any reaction.

3 If you can feel no reaction such as stinging or burning, place a small amount of the plant between your lower lip and your gum. Wait five minutes for any reaction.

4 If there is still no reaction, chew the plant · slowly but do not swallow. Hold it in your mouth for about 15 minutes.

5 If there are still no signs of irritation, then swallow the chewed piece of the plant.

6 Wait for eight hours. If you feel at all strange or if there are any side effects, make yourself vomit and drink plenty of water.

7 If there are no adverse effects, eat half a cup of the same part of the plant and wait another eight hours. If you still feel normal, the plant can be eaten normally.

▲ CHOOSING FOOD

When looking for food sources, be sure that anything you test is abundant to make the time worthwhile.

58 Insects you can eat

A common part of the diet of many indigenous jungle peoples, but outside the experience of most Westerners, insects are a valuable source of protein and are relatively easy to catch and prepare — a vital plus-point in a survival situation.

FINDING INSECTS

Insects can be found in abundance in the jungle, both above and below ground. Search for their nests in rotting logs, where you are likely to find edible beetles, grubs and termites and their larvae, or under large flat stones.

INSECT SOUP

- Place the edible insects into a preheated metal container and dry roast them over a fire for several minutes.
- Grind the cooked insects with a pestle and mortar, removing any hard legs and wings.
- Mix the powder into warm water and add edible plants to make a nutritious soup.

WHICH INSECTS?

EDIBLE INSECTS INCLUDE:

- Ants and grubs (the white ant is often considered by indigenous people to be a delicacy).
- Grasshoppers and crickets, once you have removed the wings and hard shell.

INSECTS TO AVOID INCLUDE:

- Brightly colored insects or those that have spots or stripes.
- Hairy insects such as spiders.
- Ticks, flies and mosquitoes, which may carry diseases.

▲ EDIBLE PALE WEEVIL LARVAE
Although gruesome to look at, insects and larvae are an important source of protein.

◀ COOKING INSECTS
Most insects should be cooked before being eaten — roast them over a fire.

59 Gathering shellfish

Jungle rivers are a rich source of food and, apart from fish, you can usually find a variety of freshwater crustaceans — crayfish, lobsters, crabs and shrimp.

PREPARING SHELLFISH

All shellfish must be eaten freshly caught. As they may carry parasites, it is important to boil them before they are eaten. When preparing crayfish, lobster and crabs, first remove the gills. All the flesh inside the external skeleton can then be eaten.

SHRIMP

Freshwater shrimp are normally found in the parts of a stream where the water is not running fast, either in sandy areas or near rocks. Shrimp can be attracted to the surface of a river at night and scooped off the surface of the water.

MAKING A LOBSTER POT

You can create a form of lobster pot from two pieces of bamboo. Place some bait in the pot and drop it into running water. Crustaceans enter the trap like a funnel and cannot get back out.

1 Split the bamboo and splay the shredded pieces outward in the form of a cone, leaving the base of the bamboo solid.

2 Repeat with another piece of bamboo (but remove the solid end). Place the open-ended cone inside the other.

3 Attach the two cones together with any available string, rattan or vine.

4 Attach rope or string to the wide end so that you can retrieve it for inspection.

◀ **GATHERING FISH**
Anything can be used to scoop shrimp from the water, such as a spare shirt.

60 Trapping animals

Animals are an obvious source of food in the jungle, and provide protein, which is generally lacking in plants. Special Forces often have a selection of snares in their survival kits for trapping small animals such as rodents.

PREPARING TO SET A SNARE

Identify animal runs by looking for spoor (scent markings) as well as other signs such as flattened undergrowth, animal droppings, broken spiderwebs and gnawed twigs or tree trunks. When setting a snare or trap, it is important to reduce signs of your presence as far as possible.

- Smear your hands with mud to help to disguise your body smell.
- Keep off the animal track route as much as possible and disguise any fresh breaks in branches or gashes in trees with mud or dust.

TRAPPING VERSUS HUNTING

In a survival situation, essential resources are often so limited that it is not possible to go out on extended hunting missions that may not yield any food at the end. Setting a snare, which in its simplest form takes little time, uses very little energy. Once a trap or series of traps have been set, you can lie down and rest while the traps do the work. In most cases, trapping with a snare wins over hunting as the preferred option for catching animals for food.

▼ SNARES
Vines or twigs can be used to construct traps.

SETTING A SIMPLE SNARE

A simple metal snare should be set on a known animal run and in a place where the animal will be funneled into the path of the trap, for example between stones or tree trunks. See page 83 for information on making a snare.

1 Securely anchor one end to a stake or to another firm base.

2 Position the hoop above the ground at the height of the animal's head. Support it with twigs, if necessary.

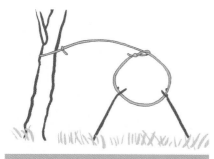

POSITIONING A SNARE

3 Once the snare is set, spread leaves and foliage around to make it look natural.

SETTING A SPRING SNARE TRAP

This form of snare uses the natural spring in a sapling to raise the captured animal off the ground, making it less likely that it will be taken by another predator.

1 Bend a sapling near the site of the snare and attach some string or a strong liana to it. Attach a hook carved from wood to the other end of the string.

2 Push an upright post into the ground (or use a tree stump) and cut a notch in it.

3 Attach the long end of the snare to the wooden hook and engage it in the notch. As the animal runs through the snare, it will also release the hook, causing the sapling to be released and spring up with the animal held in the snare.

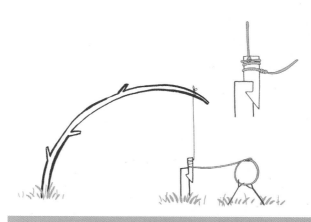

THE FINISHED SPRING SNARE SETUP

61 Making a jungle shelter

The jungle is generally a warm, covered environment and therefore a shelter is not as urgent a requirement as in extremely cold conditions. However, a shelter will boost comfort and morale, and also provides protection from tropical downpours.

CHOOSING A SITE

Special Forces will choose a site for any encampment away from well-beaten tracks, bearing in mind the need for concealment. This is less likely to be an issue for civilians in a survival situation. In this case, the following will be the main priorities:

- Choose a site for your shelter in an area that is not at risk of flash flooding from a nearby watercourse following heavy rain.
- Clear the area of rotting debris, which is likely to be infested with insects.
- Light a fire, providing there is no risk it will spread out of control, to help to keep insects at bay.

▲ BUSH CRAFT
During heavy rain, shield your campfire with collected foliage.

BUILDING A BAMBOO SHELTER

Bamboo is one of the materials most readily available in jungle areas. It also has the advantage of being light and strong, making it ideal for shelter building. You'll need at least six poles of bamboo or other wood for the frame and sufficient additional bamboo to make a roof. When cutting bamboo, take care not to splinter it — jagged edges can cause severe wounds. The black hairs on bamboo leaves are also a skin irritant.

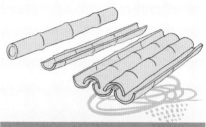

SPLIT BAMBOO TO MAKE THE ROOF

1 Split the bamboo stems for the roof from top to bottom to create slats. Interlock these slats like conventional roof tiles.

2 Use appropriate lashings (see page 187) to join the poles to make the frame of the shelter.

3 Rest the bamboo slats on the frame to create a roof. Secure if necessary with string or an improvised alternative.

4 Interweave leaves from plants such as atap (the leaves of nipa palm) through the sides of the frame to provide additional protection.

FIX THE ROOF TO THE FRAME

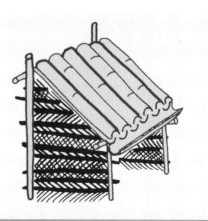

COVER THE SIDES OF THE SHELTER

62 Making a bed

A bed is not an idle luxury in the jungle – creating a raised sleeping platform is an important way of protecting yourself at night from attack by insects and from the risk of sudden flooding.

WHAT TYPE OF BED?

The type of bed you make will depend on the materials, equipment and time you have available. Instructions on how to make a bamboo pole bed are given opposite. Other options include:

HAMMOCK

A prefabricated or improvised hammock (from spare cloth) can quickly and easily be slung between two trees. This can be covered with either a poncho or a parachute.

SLEEPING PLATFORM

If you are in a swampy area and there are no trees, cut lengths of bamboo and drive them

firmly into the ground. Then tie long pieces of bamboo horizontally to form the frame and cut shorter pieces to tie on crossways.

TREE HOUSE

If you can find a suitable tree with branches that may provide the basis for building a platform, you can try building a tree house. First, cut lengths of bamboo and haul them up to the branches above to create a platform. Once you have built a stable platform, you can make a bed from leaves and use branches to provide shelter from the rain.

◀ BAMBOO VERSATILITY

Bamboo poles can be used to make shelters, platforms and even to carry water.

MAKING A POLE BED

You can use the same materials for a pole bed as for a shelter (see page 155).

1 Cut two bamboo poles to a length of about 6 feet (2 m) and lash them together to form an A-frame.

2 Lash the point of the A-frame to a tree at a suitable height. Sheer and square lashing are most suitable (see page 187).

3 Lash bamboo slats across the A-frame, starting from the longest slat holding the two splayed ends of the A-frame.

4 Push bamboo or wooden staves into the ground at the other end to provide support for the far end of the A-frame and lash them to a horizontal crossbar joining the two splayed ends of the frame.

5 Once the frame is complete, stretch a ground cloth above the bed to create a canopy.

LASH THE A-FRAME TO A TREE

LASH SLATS ACROSS THE A-FRAME

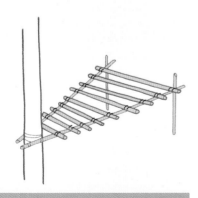

LASH THE A-FRAME TO WOODEN STAVES

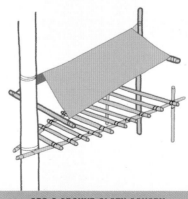

ADD A GROUND CLOTH CANOPY

63 Being snake aware

Jungles are home to many species of snake. Fortunately, many are not venomous and most will keep out of your way so long as you keep out of theirs. However, some snakes are poisonous and can inflict a painful and possibly dangerous bite, so care is essential.

DO NOT DISTURB

There can be a real danger from snakes if you disturb them or accidentally corner them — for example, when searching for fruit or cutting down bamboo or branches. You may not have time to identify the species before it attacks. Here are some tips for avoiding problems:

- **Do** avoid any snake that you encounter on a path by making a detour around it. You will be able to outrun it if necessary.
- **Do** use a stick to probe before reaching into bushes or trees — for example, when searching for food. This will give a hidden snake a warning and a chance to avoid you.
- **Do** be alert for the possible presence of snakes when moving logs — as may be necessary when building a shelter.
- **Do** be aware when walking along paths that a snake may not hear you coming and may react aggressively.
- **Do** take special care if you are moving after dark, as snakes prefer to hunt and travel at night.

ANIMAL ALERT

There are a wide variety of animals in the jungle, some of which can be hunted for food and some of which should be avoided. Dangerous animals in the jungle include big cats, such as the leopard and black panther in Africa and the Far East, and the jaguar in Central and South America. The most likely cause for attack by an animal is surprising it or accidentally cornering it. These tips will help you to avoid a potentially dangerous situation:

- Keep your distance from larger monkeys and apes as they can be aggressive.
- Beware of wild pigs, which can be extremely dangerous. The males have razor-sharp tusks and will often attack without provocation.
- Avoid game trails at night and also keep clear of water holes and rivers, where animals go to both drink and hunt.
- Don't walk through thick areas of undergrowth near rivers — you may disturb a crocodile or alligator. These are highly dangerous and will often be found near or in jungle rivers. They tend to hunt at night and are difficult to see. If attacked, you should be able to outrun one of these reptiles, although they may move surprisingly quickly to start with.

Snakes are generally well camouflaged so be careful not to disturb them when collecting logs for fires.

TREATING A SNAKE BITE

If you (or a companion) are unlucky enough to be bitten by a snake, you will need to take urgent steps in case the snake is poisonous (you may not be able to identify the snake in the incident).

1 Try not to overreact as the panic from receiving a snake bite can be dangerous in itself and also increases the heart rate, which pumps any poison around the body more quickly. If you are treating someone else who has been bitten, offer reassurance to reduce anxiety.

2 Place the bitten part of the body in a position where the bite is lower than the heart.

3 Wash the area around the bite with fresh, clean water.

4 If the bite is on an arm or leg, apply a tourniquet above the bite. Do not tie the tourniquet too tightly as this will stop all blood flow. Immobilize the limb with a splint (see page 127) and, if possible, immerse it in cold water to reduce blood flow.

5 Allow the wound to bleed; this will wash some of the venom out of the wound.

6 If possible, seek professional medical advice, unless there are signs of a good recovery or it is clear that the bite was not poisonous.

▲ STOP THE SPREAD

If the bite is on an arm, apply a tourniquet, but be careful not to tie it too tightly.

64 Stopping bleeding

Bleeding from an injury is a threat to life in any environment. In a jungle, where the use of sharp tools such as a machete is commonplace, cuts – both minor and major – are a serious risk.

IMMEDIATE ACTION

If you are away from your base when an injury occurs and have few, if any, first aid supplies, take the following immediate actions:

1 Apply pressure to any point that is bleeding using a clean cloth or bandage if available. (See opposite.)

2 If the bleeding is on a limb, raise the limb above the level of the heart.

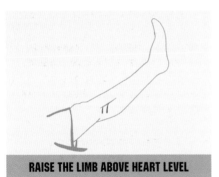

RAISE THE LIMB ABOVE HEART LEVEL

3 Use a dressing to keep the wound clean. A field dressing, consisting of a pad and bandage, is ideal. Tie a knot in the bandage away from the wound itself.

4 If blood continues to seep through the first dressing, do not remove it but place another one over it and then bandage over the dressings, keeping the bandage snug but not too tight.

CLEANING A WOUND

When you have the opportunity — for example, on your return to your base — clean the wound to reduce the risk of infection, which can have serious consequences.

1 Wash the wound with either sterile saline solution or with clean, preferably boiled water, cleaning from the center outward.

2 Thinly cover the area of the wound with antiseptic or antibiotic ointment, if available.

3 Place a sterile dressing and bandage over the wound.

APPLY A DRESSING

BANDAGE OVER THE DRESSINGS

HOW TO STOP SEVERE BLEEDING

The best and safest way to stop severe bleeding is to apply direct pressure to the wound. Alternative methods, such as applying a tourniquet above the wound, can be dangerous as they can lead to gangrene, requiring limb amputation.

CONSEQUENCES OF SEVERE BLOOD LOSS

When severe bleeding occurs, the body tries to compensate by constricting the damaged blood vessels, reducing blood flow to the skin and muscles so that it can be re-routed to vital organs such as the brain and kidneys. However, a large loss of blood will create a rapid lowering of blood pressure, which can lead to symptoms such as confusion, weakness, sweating and general pallor.

CLOSING WOUNDS

If you are treating a wound such as a knife cut, you can attempt to close the wound by using sutures or stitches. Special Forces normally carry sutures in their emergency medical kit. An alternative is to use adhesive suture strips.

USING STITCHED SUTURES

1 Clean the wound as described opposite.
2 Using a sterilized needle (heat the needle over a flame before use), pass it through one side of the wound to the other.
3 Use as many sutures as necessary to close the wound, tying off each one individually.

▼ KEEP SHARP

A sharper machete is less likely to slip and cause injuries.

ADHESIVE SUTURE STRIPS

1 Clean the wound as described opposite and pull the edges together.
2 Stick one side of the adhesive strip to one edge of the wound.
3 Keeping the wound as firmly closed as possible, stick the other end of the strip firmly to the other edge of the wound.
4 Apply additional strips in the same way if the wound is large.

In all cases, cover the area with a clean dressing after suturing (see opposite).

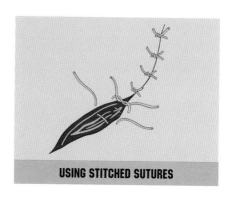

USING STITCHED SUTURES

65 Finding your way in the jungle

The density of the jungle canopy can make it difficult to gain a sense of direction either through the movement of the sun or the position of the stars. It is also often impossible to take a bearing on a distant landmark, as you would in more open country.

WHAT IS DEAD RECKONING?

Dead reckoning is a useful way of maintaining a sense of direction when visibility is restricted — as in the jungle. This method of navigation can help you progress toward a known destination if you are aware of your current position, through the use of compass bearings and careful calculation of the distance traveled. You can make an accurate estimate of the distance traveled by working out the distance you cover in a set period of time in different conditions — for example, on flat ground or uphill. You will need a compass and map, a watch and a means of keeping a log, such as a notebook.

HOW TO NAVIGATE BY DEAD RECKONING

1 Measure or estimate your walking speed both on flat ground and when going uphill (which will be slower). An average walking pace on flat ground is 3 miles per hour (about 5 km/h), meaning that it will take about 1 minute to cover 300 feet (about 90 m).

2 Check the time and the compass direction (course) in which you are traveling and note both in your log.

3 Note the time when you reach a point at which you need to change direction.

4 Note the new compass direction in which you are traveling in your log and start counting the paces again.

5 Repeat this process each time you change direction (see example below).

6 Draw your route on your map to ensure you are maintaining the desired direction.

Time	Course	Distance
0800	090	3 miles (5 km)
0900	035	1 mile (1.5 km)
1130	090	2.5 miles (4 km)

▲ ESSENTIAL TOOL
A compass is an essential tool in the jungle.

CHOOSING A ROUTE

In some areas, it may be advisable to follow a ridge in order to make the going easier. You will need to make an estimate of how close the ridge is to your intended direction of travel versus the ease of movement.

In the jungle there may also be a variety of trails, created by either animals or humans. The animal trails may lead to or from water. Paths made by humans may lead deep into the jungle — such trails are often created by loggers searching for suitable trees.

◀ FIND THE TRAIL

U.S. Marines Corps move cautiously along a narrow jungle trail while on patrol as a part of jungle warfare training in Colombia.

JUNGLE VISION

Special Forces are trained to move quietly and inconspicuously through the jungle, often pausing to listen and to take their bearings. If you are stranded in the jungle you can learn from these techniques.

- Look through the jungle rather than at the foliage immediately in front to see into the distance.
- Stop regularly and lower yourself to the ground in order to look along the jungle floor and identify any useful animal or human tracks.

MOVING INCONSPICUOUSLY

66 Cutting through the jungle

In order to travel through dense jungle in which there is unlikely to be any kind of path, it is usually necessary to cut your way through the vegetation that blocks your route. This can be laborious, but doing it efficiently can speed your journey to safety.

CUTTING YOUR WAY THROUGH

Traveling through dense vegetation is slow — especially when you have to cut a path. You may achieve no more than about half a mile (1 km) an hour, which could mean a maximum distance of about 3 miles (5 km) a day. Special Forces use the following techniques when moving through dense jungle:

- Use a machete to cut through thick jungle vegetation.
- Use an upward stroke when cutting through vines and other thick vegetation — it is more effective.
- Use a stick to part thin-stemmed vegetation in front of you. The stick will also move any unpleasant insects or reptiles out of your path, saving you from a potentially dangerous encounter.

FINDING YOUR WAY BACK

If you intend to return to camp, mark your trail so that you can easily find your way back. Use any of the following methods:

- Cut slashes in trees.
- Cut palm leaves and turn them upside down to show their light undersides.
- Use rocks or make marks in the trail.

AVOID TRAVELING AT NIGHT

Due to the density of vegetation, the difficulty of seeing your way in the dark and the possibility of coming across dangerous animals, it is best to avoid moving at night in the jungle.

◀ LEAVE A TRAIL
A U.S. Marine marks a path through the jungle using strips of palm leaf.

▶ SAFE CROSSING
A basic raft in progress, made from bamboo poles.

67 Traveling by water

Because of the dense vegetation and difficulty in navigation, Special Forces often judge traveling by river to be the best option in the jungle. However, even if you have the skills and materials to build a raft, it can also present serious difficulties and dangers.

KNOW THE RISKS

While river travel may be quicker than walking through the jungle on land, it carries significant dangers that you need to take into account.

- Extreme pressure on a raft due to the movement of the water can weaken lashings.
- Rapids and rocks could smash the raft or fatally weaken it.
- Jungle rivers can contain dangerous animals such as alligators.

WHY TRAVEL BY RIVER?

One reason to travel by water is that it often offers the best chance of meeting local people from whom you may be able to seek help. Villages are normally situated on the banks or at the confluence of rivers. Remember that when approaching villages, it's best to wait outside until you are invited in.

BUILDING A BASIC RAFT

Special Forces use this method to make a basic raft to use for traveling down a jungle river. You'll need sufficient thick bamboo stems to make two layers of the width of raft you require, rope or tough vines to make the lashings and some small sticks to use as pegs.

1 Cut the bamboo stems into approximately 10-foot (3 m) lengths.
2 Make holes through each of the lengths of bamboo at both ends and also halfway along their length.
3 Push small wooden sticks through the holes and use a round lashing to secure them to the bamboo.
4 Repeat the process with a second layer of bamboo, resting it on top of the first layer.
5 Lash two bamboo stems together across the ends in order to lock the two layers of bamboo together.

MAKING A LOG RAFT

You can create a similar raft with logs. These may be thick enough to provide sufficient flotation with one layer. In this case, place crossbars across the logs on the top and bottom, lashing them together at each end to create sufficient pressure to hold the logs in place. Make sure that the logs will not move out from under the crossbars due to the stress created by movement on the water.

▶ CHOOSE LOGS WITH CARE
It is better to recut a pole than try to make do later.

A-FRAME STEERING SYSTEM

You can create a steering system for your raft by building an A-frame from bamboo or other wood using the method described on page 157.

1 Once you have constructed an A-frame, place the two splayed ends in holes that you have cut on each side of the raft.

2 Run a rope from the top of the A-frame above the lashing and tie it securely at the front and back on each side of the raft.

3 Take a long pole and create a rudder with whatever materials you may have available, such as a square plank. Attach the pole to the A-frame with a sheer lashing that allows sufficient movement for steering.

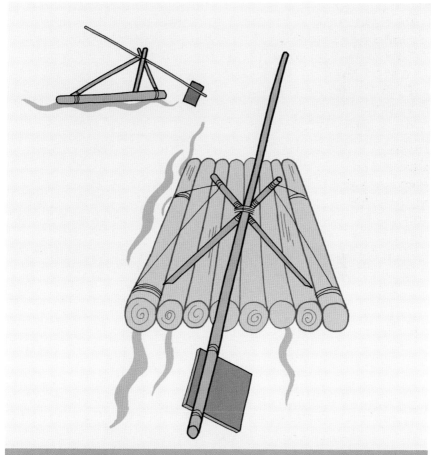

HOLD THE RUDDER AGAINST THE FLOW OF WATER TO ALTER DIRECTION

68 How to cross a river

If a river lies across the route you want to take, you may have to cross it. Building a raft may be advisable if the river is very wide (see pages 170–171), but in other cases — bearing in mind hazards such as currents and rapids — a crossing through the water may be best.

CHOOSING YOUR CROSSING POINT

Make sure that the point where you intend to cross the river is easily accessible from your side of the bank and that you will be able to get out easily on the other side. Take into account the fact that you are likely to be getting out of the river at an angle to where you go in, depending on the direction and strength of the current.

- Do not cross at a bend in the river, as the water tends to run more rapidly here.
- Beware of rocky areas, which may be slippery underfoot.

- When crossing a river, it is a good idea to take a strong stick or pole with you to feel ahead for secure footings and also to provide stability. Keep the pole upstream of your body.
- In order to keep your clothing dry, put your pants, shirt and coat in a pack worn high on your back that you can remove rapidly if you are swept away by the current — it may drag you underwater.
- Do not remove your footwear as it will provide you with better grip than bare feet, as well as protection.

◀ ROPE GUIDE
A rope is a useful aid when crossing a deeper river with a strong current.

THREE-MAN RIVER CROSSING

You can use this method if there are three or more people in the group and you have a rope of sufficient length to make a loop that spans the river.

1 Having made a loop in the rope, the strongest person wades into the water holding the rope in front of them so that they are within the loop. The others on the bank hold the rope loop securely.

2 After reaching the far bank, the first person holds or secures the rope, while the second strongest person continues to hold the rope on the near bank. Repeat this procedure for all weaker members of the group until only one person is left on the near bank.

3 The last person then crosses the water within the loop, secured by the team members on the far bank.

FIRST PERSON CROSSES IN THE LOOP

SECOND PERSON CROSSES

LAST PERSON CROSSES

69 Making a raft for river crossing

You can create a simple raft from a poncho liner or from a few logs to use for crossing a river. Using this technique could save you many days of laborious travel to a ford or other crossing point.

MAKING A WOOD SEAT RAFT

This improvised raft is relatively easy to construct and will provide basic flotation, helping you to keep most of your body dry. It has the disadvantage of being difficult to control and is only suitable for short crossings of relatively slow-flowing rivers.

1 Cut two large logs to a length of about 4 feet (1.2 m) and lay them parallel.

2 Place the logs so that you are able to sit comfortably between them with your legs draped over one and your back resting against the other. Try sitting between the logs on dry land to find the most comfortable position before deciding on the gap between them.

3 When you have determined the correct gap between the logs, tie the logs together with rope, allowing sufficient length to enable you to sit comfortably between.

D.I.Y. FLOTATION AIDS

Any object that can trap and hold air, such as plastic bottles or empty fuel cans, can in theory be used as a makeshift buoyancy aid to help you to get across a river with greater safety if it is too wide or too fast to swim across with confidence.

◀ WOOD FLOATS
Any large log can be used as a float to cross a river.

MAKING A PONCHO RAFT

This type of raft is suitable for a short river crossing, but is unlikely to withstand a longer journey. You'll need a poncho or other piece of waterproof fabric, some sticks and sufficient dry foliage to provide stuffing.

1 Place two sticks in a cross on your open poncho.

2 Place light, dry foliage over and around the crossed sticks.

3 Wrap the poncho around the foliage and tie it securely.

4 If possible, wrap a second layer of material around the first to keep the raft watertight.

FOLIAGE AROUND CROSSED STICKS

ADD MORE DRY FOLIAGE

WRAP AND TIE THE PONCHO

WRAP A SECOND LAYER

5 URBAN SURVIVAL

At times of natural disaster when civil order may have broken down, as well as during armed conflict, towns and cities can easily become hostile areas in which those caught up in events can be at serious risk. Special Forces use specialist techniques for survival in urban areas — often to avoid contact and blend in.

SAFE ON THE STREETS

An urban area may not seem like the type of environment in which you would need to use survival skills, but during a natural disaster or civil emergency, the knowledge of specialist techniques could be essential for your survival.

We are used to thinking of urban areas as places where the essentials of life are easily obtained. However, in times of civil unrest or following catastrophic destruction — in the wake of an earthquake, for example — such areas can easily become threatening.

As warfare increasingly goes beyond battles between conventional armed forces, anyone can become involved in a combat situation — perhaps during a terrorist incident, if taken hostage, or as a result of the threat of a bomb attack. Hints from the Special Forces on how to stay alert may be lifesaving for anyone caught up in this kind of situation.

Earthquakes, floods, hurricanes and fires can affect urban areas as often as any other environment. Special Forces learn to cope with these kinds of natural emergencies through intensive training so that they can master any situation, however unpredictable it may be.

SURVIVAL CHALLENGES

Obtaining the basic survival essentials of shelter, food and water may not present a major problem in an urban area, unless there has been a natural disaster such as an earthquake or tsunami. The challenges in urban areas are often more concerned with

dealing with the threat from hostile or criminal elements of the population. For civilians, the most likely reason for the need for survival skills is in the case of a sudden deterioration in the political situation and an increase in hostility. In such cases, both physical and psychological survival skills may be required.

Special Forces spend a considerable amount of time training for combat in built-up areas. Urban environments provide some of the greatest challenges to their skills, considering that terrorists and insurgents can easily blend in with the local population. Civilians stranded in a foreign city in which a proportion of the residents may be hostile can benefit from such knowledge.

URBAN EQUIPMENT

In terms of preparation and equipment, clothing is not usually an important issue due to the ready availability of shelter. Simply dressing appropriately for the climate of the region is usually sufficient. Special Forces nevertheless carry an emergency survival kit, which includes: a reliable flashlight, whistle (for calling for rescue), notebook and pen, needle, thread and safety pins (for emergency repairs), a basic first aid kit and a multi-tool, such as a Swiss Army knife.

Another aspect of urban preparedness is the consideration of transport. If you are

▲ URBAN TERRAIN
Special Forces soldiers were deployed into urban areas in Iraq and Afghanistan.

◀ CLEARING HOUSES
Soldiers practice room-clearing procedures in a tactical entry house for use in securing buildings.

traveling by car, make sure that the vehicle is well maintained and that you are carrying commonly needed spares: tires, fuses, lights, jump leads and basic tools. Don't rely on satellite navigation systems for finding your way — these can fail. Always carry a conventional map as a backup.

Urban areas are full of noises and activity that demand attention as potential sources of danger, including crowds, traffic congestion and a multitude of visual distractions, such as shop windows and advertising billboards. This chapter is about how to fine-tune the kind of instincts that Special Forces use to keep safe when operating in an urban environment, and how these skills can be adapted by anyone who finds themselves in threatening conditions.

SURVIVAL SITUATIONS

Those operating on missions in built-up areas often find themselves in situations where high levels of awareness are essential in order to evade pursuers and, if necessary, to act in self-defense. On an urban mission it is also important to be familiar with techniques such as how to travel safely — whether on public transport, by car or when walking the streets. Buildings may no longer offer the possibility of safe shelter, but can become a source of danger — either because of physical instability or because of a threat from hostile inhabitants.

Familiarity with Special Forces principles can help you to become more aware, better prepared and more confident to take on even the most unexpected survival challenges in built-up areas.

70 Traveling in hostile territory

Special Forces learn to keep away from potentially risky situations as they move around urban areas, and take care to avoid attracting unwanted attention.

CHECK THE AREA

In a hostile urban area where kidnapping or robberies are known to happen, Special Forces are trained to take the following precautions:

- When exiting a building, look up and down the street to check for suspicious people either loitering in the street or in cars.
- Be particularly wary if you spot two or more men in a parked car.
- When walking down the street, walk on the side of the street that allows you to face oncoming traffic.
- Keep away from the side of the road and be aware of any cars that may be following you slowly or that suddenly slow down. If this should happen, run quickly toward other people to find safety in a crowd.
- Keep clear of dark alleys and other potential hiding places for attackers.

CAR SAFETY

If possible, drive a local car with local license plates. When driving in a high-risk area, be aware of ruses that may be employed to tempt you to slow down or stop. These may include: a fake request for a lift or directions, a hoax cry for help by a pedestrian or a staged accident.

SAFETY ON PUBLIC TRANSPORT

When traveling on trains or buses, Special Forces are trained to take a seat near an exit. The following precautions ensure that it is easy to get out quickly and avoid a potentially dangerous situation.

- Sit near the caboose of a train.
- Stand some distance away from the line to board the bus, until it arrives, and then board quickly.
- Avoid falling asleep while traveling, as this leaves you vulnerable.

▲ SWAT TRAINED

Special Forces troops have techniques for checking an area is safe and clear of hazards.

71 Going unnoticed

It is important to move covertly in a hostile urban environment. You should aim to blend in as much as possible to avoid drawing attention to yourself – going unnoticed may be the key to survival.

BLEND IN

If you do not wish to be identified as an outsider in a foreign city, you need to ensure that there is nothing about your movements, actions or clothing that attracts undue attention. Special Forces follow these principles to blend in:

- Wear and carry similar clothing and accessories as local people. For example, do not carry a leather briefcase if everyone else is using backpacks.
- Behave like the people around you. If you are on a busy street where most people are going to work, walk in the same purposeful manner. If you are in a place where most people are relaxing or shopping, adapt your behavior and movements in a similar way.

AVOIDING DETECTION

An urban environment provides numerous hiding places for an enemy to conceal themselves or to observe your movements. Special Forces use the following techniques to avoid being observed.

- Avoid passing windows. Instead, duck your head or, if necessary, crawl beneath them. If there is a basement window, step over the area of the window frame.
- If you want to look around a street corner where you think there may be an enemy presence, get down on your stomach and cautiously look around the corner at ground height.
- Stay close to the wall when moving up a sidewalk.
- Before making a move, check if there are any places you can use to conceal yourself temporarily. If it is daylight, look out for areas in the shade.

BEWARE OPEN SPACES

Keep away from open areas, such as parks, where you may not be able to avoid being seen.

◀ STAYING HIDDEN
A Marine checks the area from a low height during training.

72 Evading capture

Special Forces are trained to take steps to prevent capture as a hostage in a foreign environment. If a governing regime is unstable, you may have to avoid the authorities too.

DETERRING CAPTURE

One of the best ways of deterring planned capture is to avoid following a set routine. Here are some routine actions to avoid:

- **Don't** go for regular walks or runs along the same route every day.
- **Don't** use the same route for going to or returning from work every day.
- **Don't** have a regular day for shopping or always use the same supermarket.
- **Don't** frequent the same bars or restaurants regularly.

ARE YOU BEING FOLLOWED?

If you think that you are being followed, try these techniques for testing your suspicions:

- Keep an eye out for anyone who does not seem to have anything in particular to do or who repeatedly looks in your direction.
- Try walking into a shop and looking through the window to see if the person walks past or glances in to check if you are inside.

TRUST A POLICE OFFICER?

If you are in a nation whose political stance is against that of your native country, seeking help from the local police may not be the right solution and could get you into further trouble. If that is the case, seek help from your national consulate or embassy or the officials of a country you know are friendly to yours.

SHAKING OFF A PURSUER

If you are being pursued, move to a safe place such as a police station as soon as possible, or move to a public place such as a hotel foyer.

You can try to lose your pursuers by taking public transport, but be aware that someone may follow you. If you think you are being watched, get off at the next opportunity and see if the suspected pursuers follow. Stay vigilant — another person may be waiting to pick up the surveillance at the next stop.

◀ LOOKING OUT FOR YOURSELF
In unknown territory it can be difficult to know who to trust.

▶ HIDDEN FLAMES
Special Forces sheltering in an abandoned building need to be sure that a fire won't give away their position.

73 Setting up shelter

Special Forces are trained to find suitable places in urban environments from which to carry out reconnaissance. The same principles can be used to shelter out of sight in any urban area.

CHOOSING THE LOCATION

Special Forces select a location where people do not regularly come and go. It is also important to choose a site that blends easily with the surrounding area and does not attract attention. Sites often used by Special Forces include abandoned factories and attics in abandoned buildings. Here are some tips to bear in mind:

- Avoid the obvious. If you choose the first site that seems suitable to you, ask yourself whether an enemy would think the same thing.
- Make sure that there is a second escape route from the site in case you are discovered.
- Take care not to use a building in such a deteriorated state that there is a risk of structural collapse.

SAFETY AND SECURITY

When you move in to your chosen site, you need to check that it is safe. Special Forces check for dangers such as booby traps and they also take care in areas where they might be seen through windows or external holes in the building. You will need to identify an area that cannot be seen from the outside if you need to use lights or light a fire. To avoid detection use the following techniques:

- **Avoid creating visible smoke** — If you cook, check that you do not produce smoke that can be seen from outside the building.
- **Beware of signs of heat** — If you are producing heat inside a building and it is cold with ice or snow outside, you might betray your position by melting the snow and ice on the building exterior that you are occupying.
- **Avoid any unusual coming and going** — If you are in an abandoned site you will need to take care that you do not leave obvious footprints leading to and from the site.
- **Don't make obvious changes** — If you have had to make any alterations in preparing your shelter, make sure that you leave everything undisturbed on the outside of the building to avoid attracting unwanted attention from passersby.

74 Defending yourself

If you have no self-defense training, your best option is to avoid dangerous situations, if necessary outrunning an attacker or seeking help from others.

SURVIVING A KNIFE ATTACK

If escape isn't an option, try the following Special Forces techniques:

- Use a weapon such as a stick to hit the knife out of your assailant's hand. Keep the stick out of reach of your attacker so that he cannot grab it instead.
- If you have no weapon and your attacker makes a lunge toward you, keep your upper body back and out of the way while kicking out with your legs at his shins or knee caps. This may cause enough disabling pain to distract the assailant while you get away.
- Wrap a coat or any other available thick material around your forearm so that you can parry the knife arm.

DEFENSE FROM A STICK OR CLUB

When dealing with this threat, Special Forces are taught — in contrast to dealing with a knife attack — that getting close to the assailant reduces the impact of the attack. They close in as soon as possible and retaliate decisively to disarm the assailant.

▼ SPECIAL SKILLS

At an advanced level, Special Forces perform martial arts training to defeat attackers.

SELF-DEFENSE BASICS

There are certain principles and techniques that may improve your chances in a fight, and specific places that you can hit your assailant to cause maximum distraction and give you those vital moments to get away.

- Maintain a guard position by keeping your fists near your face to parry blows.
- Fling grit or sand in your assailant's eyes to temporarily disable him. Alternatively, poke his eyes with your fingers.
- Attempt to land a blow on your assailant's nose, which can cause a temporary concussion.
- Aim a punch in the solar plexus area of the chest just under the breastbone to wind your assailant.
- Kick or hit your assailant's shins, knees or elbows; this can cause disproportionate pain and may be temporarily disabling.
- Aim a kick to the groin. This is certain to concentrate your assailant's mind.
- Stamp on the top of your assailant's foot. This is painful and may impair his ability to pursue you.

▲ PROTECT YOURSELF

Maintain a guard position to protect your face during an attack.

6 ESSENTIAL KNOTS

A skill required for every survival scenario, knot tying is a core technique that Special Forces are required to master. Knots are used by Special Forces to build shelters, create snares and traps, make rafts and improvise stretchers. Rope work and knot tying are also crucial for many mountaineering techniques.

STOPPER KNOTS

Stoppers are used as a brake to prevent a rope from running freely through a hole or fixture.

OVERHAND KNOT

This knot can be used to secure the end of single or double small-diameter cord, line, or even sewing thread — either as a small stopped knot or to prevent the ends from unraveling.

FIGURE EIGHT STOPPER

This is perhaps the most commonly used stopper knot. It is more bulky than the overhand knot (above) and easier to untie. However, with excessive shaking, this knot may loosen or even become undone. A variation on this knot creates a standing loop (see page 185).

OVERHAND KNOT

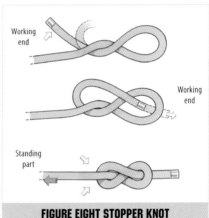

Working end

Working end

Standing part

FIGURE EIGHT STOPPER KNOT

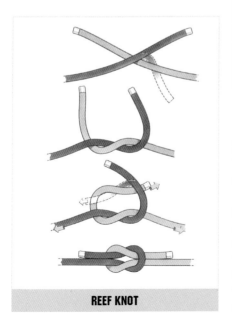

REEF KNOT

BINDING KNOTS

These knots are usually used to secure the ends of a rope around an object.

REEF KNOT

Also known as a square knot, this knot is designed to join two ends of a rope of equal thickness. It is reliable under strain but less so when loose. It is useful as a knot for securing the ends of a bandage as it lies flat and is therefore more comfortable.

BENDS

Bends are used to join two pieces of rope — something that is likely to be needed in a survival situation when improvisation is often essential.

SHEET BEND

Also known as the common bend, the sheet bend is used to join two lines of different thickness (provided the difference is not too great). A sheet bend can also be used to secure a rope to anything that has a hole through which the line can be passed and the working end then trapped under the standing part — for instance, with a hammock ring. Another common use is in the knitting or repair of netting, such as for fishing.

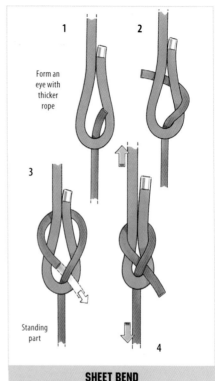

1

2

Form an eye with thicker rope

3

Standing part

4

SHEET BEND

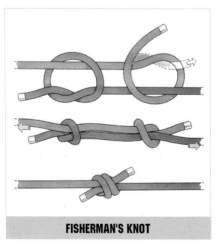

FISHERMAN'S KNOT

FISHERMAN'S KNOT

This knot is formed by drawing together two overhand knots (page 183) so that the short, working ends lie in opposite directions. This knot can be used to join rope and other natural materials such as vines. It is very secure and is difficult to untie.

LOOP KNOTS

These knots are used to create a fixed loop in the end of a rope — as may be needed when rescuing someone from the water, for example (see page 91).

BOWLINE

This is the most commonly used loop knot. It is strong and stable. Its possible uses include securing a line to a harness or ring, as a handhold at the end of a line or when fixing a rope to an anchor point (see page 119).

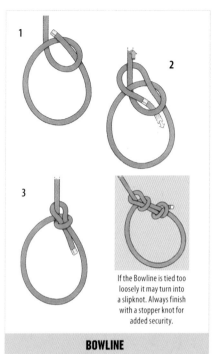

If the Bowline is tied too loosely it may turn into a slipknot. Always finish with a stopper knot for added security.

BOWLINE

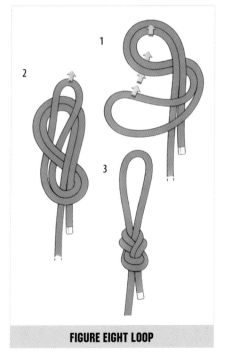

FIGURE EIGHT LOOP

FIGURE EIGHT LOOP

A figure eight loop is a reliable fixed loop that is easy to remember and tie. It has the advantage that it can be tied in the bight of a rope if the two ends are not accessible, or in the end of a rope which has been doubled to make a bight. When tying, pass the bight around the standing part(s), then pull the knot up so that the leads lay neatly and parallel alongside each other.

HITCHES

Knots in this category are used to attach a rope to a rail or other apparatus or equipment.

CLOVE HITCH

This is a useful knot for attaching a rope to a pole or a tree. It can be used in climbing, for example, where a rope is attached to a carabiner. It should be used with caution, however, as it may come undone if the object around which it is tied can rotate, or if constant pressure is not maintained on one or both ends of the rope.

TIMBER HITCH

The timber hitch is quick to tie, never jams and is easy to untie. It is useful for securing heavy objects such as logs or branches that you want to drag. Be sure to pass a fairly long working end around the object, take it around the standing part and then take the working end back around itself at least three times.

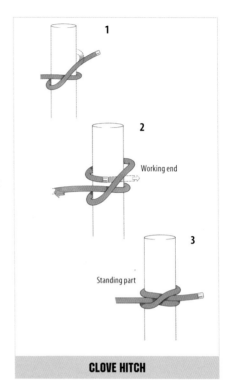

CLOVE HITCH

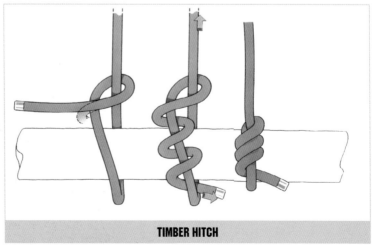

TIMBER HITCH

LASHINGS

Lashings are ropes or vines used for tying, wrapping or connecting objects together, such as bamboo used for making a raft or shelter.

SQUARE LASHING

The square lashing is used to lash together two poles at right angles. This is one of the most useful knots for building shelters, platforms, rafts and other items necessary for survival. Start with a clove hitch (opposite) around the upright, twisting the tail around the line (Step 1). Make sure that each turn is pulled up tight before the next is applied. Three or four turns should be sufficient, with two or three frapping turns (the turns around a lashing that tighten it before it is secured). Finish off with a clove hitch, started as close to the frapping turns as possible.

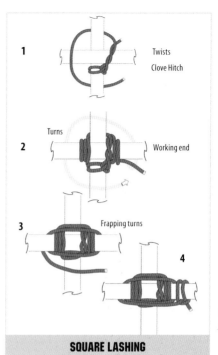

1 Twists / Clove Hitch

2 Turns / Working end

3 Frapping turns

4

SQUARE LASHING

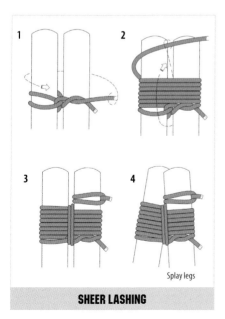

1 **2**

3 **4**

Splay legs

SHEER LASHING

SHEER LASHING

Sheer lashings can be used either to join two poles so that the legs can be splayed or to lash two poles parallel to each other. This lashing may be used, for example, to form an A-frame for the steering frame of a raft (page 167). The width of the lashing should be at least the combined width of the two poles. Start and finish the lashing with a clove hitch (opposite). When extending the length of a pole, use two sheer lashings without frapping turns. Tighten each turn by hand as much as possible, then hammer wedges between the poles and the lashing to tighten further.

Index

Page numbers in italics refer to illustrations

ACKNOWLEDGMENTS

The publisher wishes to thank:
For editorial: Lucy Kingett, Cathy Meeus, Etty Payne. Indexing: Diana LeCore. For illustration: Jess Wilson. For design: Paul Turner. Additional thanks to Gareth Butterworth and Sue Pressley.

Author Alexander Stilwell would like to thank:
Sorrel Wood, editorial director at Marshall Editions, for asking me to write this book and for her encouragement; Philippa Davis at Marshall Editions for her support and guidance; and Cathy Meeus for her patient and persevering editorial backup, worthy of the Special Forces.

Wikimedia Commons; page 41, 45 United States Navy SEALs/ Wikimedia Commons; page 42 Andrey Grinyov/Shutterstock. com; page 44l Pavel Vakhrushev/ Shutterstock.com; page 44r, 46t SurangaSL/Shutterstock.com; page 46b U.S. Air Force/Tech. Sgt. Jeremy T. Lock; page 47t Crown Copyright/Royal Navy/ POA(Phot) Dave Husbands; page 47b Steve Byland/Shutterstock. com; page 48 U.S. Navy/ Mass Communication Specialist 1st Class Roger S. Duncan; page 49bck Vladimir Meinikov/ Shutterstock.com; page 49, 52 U.S. Army Alaska/Sgt. Patricia McMurphy/dvidshub; page 50 U.S. Navy/Petty Officer 1st Class Roger S. Duncan; page 51, 68 U.S. Air Force/Senior Airman Cynthia Spalding; page 53t Crown Copyright/Royal Navy/ LA(Phot) Dave Jenkins; page 53b Tyler Olson/Shutterstock.com; page 58 Premysi/Shutterstock. com; page 59 OlegDoroshin/ Shutterstock.com; page 60 Crown Copyright/Royal Navy/PO(Phot) Mez Merrill; page 61 U.S. Marine Corps/Lance Cpl. Ali Azimi; page 62 Mikadun/Shutterstock.com; page 63 U.S. Marine Corps/Cpl. Sarah Dietz/dvidshub; page 64 Crown Copyright/MoD/Royal Navy/POA(Phot) Terry Seward;

page 66 U.S. Marine Corps/ Lance Cpl. Kelsey J. Green; page 69 U.S. Marine Corps/Lance Cpl. Suzanna Lapi/dvidshub; page 70 U.S. Air Force/Senior Airman Jonathan Snyder; page 74 Kirsanov Valeriy Vladimirovich/ Shutterstock.com; page 75 Robert Hoetink/Shutterstock.com; page 76 Maksimilian/Shutterstock. com; page 81b Vincent Drolet Lamarre/Shutterstock.com; page 82 Incredible Arctic/Shutterstock. com; page 83 Alan Scheer/ Shutterstock.com; page 84t BMJ/ Shutterstock.com; page 84bl Kim Hansen/Wikimedia Commons; page 84br Madien/Shutterstock. com; page 85tl Taina Sohiman/ Shutterstock.com; page 85tr JeniFoto/Shutterstock.com; page 85bl Rusian Kudrin/Shutterstock. com; page 85br Trofimov Denis/ Shutterstock.com; page 86 U.S. Air Force/Justin Connaher; page 87 U.S. Marine Corps/Cpl. Nicole A. LaVine; page 88 Jomegat/ Wikimedia Commons; page 90 U.S. Army/Staff Sgt. Brehl Garza; page 92 Canadian Land Forces Command/Cpl. Laviolette; page 94 4th Brigade Combat Team, 1st Infantry Division/dvidshub; page 95bck Tomas Tichy/Shutterstock. com; page 95m U.S. Marine Corps/Cpl. William Jackson; page 96 U.S. Marine Corps/LCPL

Stephen Kwietniak; page 97, 182m U.S. Army/John D. Helms; page 98 U.S. Marine Corps/LCPL Chad J. Pulliam; page 99l U.S. Air Force/Staff Sgt. Christopher Ruano; page 99r U.S. Marine Corps/Lance Cpl. Carlos Sanchez Valencia; page 101t U.S. Marine Corps/Lance Cpl. Sean Dennison; page 101b Crown Copyright/ Defense Imagery/POA(PHOT) Mez Merrill; page 104 Smit/ Shutterstock.com; page 105 Sergey Novikov/Shutterstock. com; page 106 U.S. Marine Corps/ Cpl. Andrew D. Thorburn; page 108 dahu 1/Wikimedia Commons; page 109 DMSU/Shutterstock. com; page 111 Daniel Prudek/ Shutterstock.com; page 112 elwynn/Shutterstock.com; page 113 Alexander Ishchenko/ Shutterstock.com; page 114, 115 Ed Darack/Getty Images; page 117 U.S. Marine Corps/ Sgt. Ben J. Flores; page 119 USAJFKSWCS/Flickr; page 121 U.S. Marine Corps/Lance Corporal Andrew J. Carlson; page 122 Sihasakprachum/Shutterstock. com; page 124 Jef132/Wikimedia Commons; page 126 U.S. Marine Corps/Wikimedia Commons; page 128, 130, 131t, 132, 133l, 135, 138, 151, 161, 163b, 164, 168, 170 U.S. Marine Corps/Cpl. Brian J. Slaght; page 129bck Boleslaw

Kubica/Shutterstock.com; page 129m Maridav/Shutterstock. com; page 131b, 147br Robert Adrian Hillman/Shutterstock. com; page 133r, 140c, 141tl Dr. Morley Read/Shutterstock.com; page 136 Juhku/Shutterstock. com; page 139 SSGT Tony Bauer/ Defense Imagery; page 140l Kletr/Shutterstock.com; page 140r Schubbel/Shutterstock.com; page 141tr Shaiith/Shutterstock. com; page 141cl Hendroh/ Shutterstock.com; page 141b mrfiza/Shutterstock.com; page 143, 156, 166 SSGT Marvin Lynchard/Defense Imagery; page 144l Steve Heap/Shutterstock. com; page 144r K. Luzzi Paul; page 145tr Anneli Salo/Wikimedia Commons; page 145bl JPL Designs/Shutterstock.com; page 145br Wikimedia Commons; page 146l Sommai/Shutterstock. com; page 146r MajiPineapple/ Shutterstock.com; page 147tl kowit sitthi/Shutterstock.com; page 147tr Dobino/Shutterstock. com; page 147bl Lisa F. Young/ Shutterstock.com; page 148tl Casper1774 Studio/Shutterstock. com; page 148tr H. Zell/ Wikimedia Commons; page 148bl freedomman/Shutterstock. com; page 148br jo Crebbin/ Shutterstock.com; page 149 U.S. Department of Defense/U.S.

Marine Corps/ Sgt. Benjamin E. Barr, USMC; page 150l U.S. Marine Corps/Cpl. Jonathan G. Wright; page 150r Lukasz Janyst/ Shutterstock.com; page 152 Marine Expeditionary Force/ Lance Cpl. Joseph Cabrera; page 154 Marine Expeditionary Force/ Cpl. Paul Zellner; page 159t mroz/ Shutterstock.com; page 159b U.S. Air Force/Senior Airman Kamaile O. Chan; page 162 U.S. Air Force/ Staff Sgt. Jacob N. Bailey; page 163t U.S. Marine Corps/Lance Cpl. Ammon W. Carter; page 165 Valery Shanin; page 172 U.S. Army/Sgt. 1st Class Johancharles Van Boers; page 173bck Oliver Sved/Shutterstock.com; page 173m U.S Marine Corps/Lance Cpl. Alfredo V. Ferrer; page 174 U.S. Air Force/Senior Airman Sam Goodman; page 175 U.S. Marine Corps/Lance Cpl. James J. Vooris; page 176, 177 U.S. Marine Corps/Lance Cpl. Joshua J. Hines; page 178 Sgt. Pete Thibodeau/ dvidshub; page 179 U.S. Air Force/Staff Sgt. Clay Lancaster; page 180 U.S. Marine Corps/ Lance Cpl. Claire A. Prindable; page 181 U.S. Marine Corps/ Cpl. Tommy Huynh; page 182bck MILpictures by Tom Weber/Getty Images.

All other images are the copyright of Marshall Editions. While every effort has been made to credit contributors, Marshall Editions would like to apologize should there have been any omissions or errors and would be pleased to make the appropriate correction to future editions of the book.